DESIGNING VALID RESEARCH

A Brief Study of Research Methodology

Hollis L. Green, ThD, PhD

GlobalEdAdvance
Press

DESIGNING VALID RESEARCH
Copyright © 2011 by Hollis L. Green

Library of Congress Control Number: 2011923855
Green, Hollis L., 1933 –
    Designing Valid Research: A Brief Study of Research Methodology
    ISBN 978-1-935434-57-3

    Subject Codes and Description:
    1: EDU.037000: Education: Research 2; SCI 043000: Science:
    Research and Methodology 3; EDU 008000. Education: Decision
    Making and Problem Solving

Cover Design by Barton Green
Author's Photograph by Carie Thompson: cariephoto@gmail.com

Published by
GlobalEdAdvancePress
**gea-books.com or
GlobalEdAdvance.org**

This book is respectfully dedicated to

G. A. SWANSON, PHD
(1939 - 2009)

Dr. G.A. Swanson was an American organizational theorist, and a tenured Professor of Accounting and Business at Tennessee Technological University (1982-2009) He was known for his accounting theories based on James Grier Miller's general living systems theory (LST) and his passion for social scientific research. He served on editorial boards of Systems Research and Behavioral Science, Systems-Journal of Trans-disciplinary Systems Science, Journal for Information Systems and Systems Approach and International Encyclopedia of Systems and Cybernetics. Swanson was president of International Society for the Systems Sciences (1997).

Dr. Swanson was awarded an Institute of Internal Auditors Research Foundation Fellowship (1989-90), a D.Litt. (1991) at the Oxford Graduate School, a College of Business Administration Excellence in Overall Performance Award (1997), and a College of Business Administration Foundation Award for Outstanding Research (1987, 1993, and 2004).

As the author of many texts and Journal articles, together with his passion for research in the integration of religion and society, he was an inspiration to me. His wise counsel through the years was appreciated and most meaningful. It was my privilege to co-author two texts with Dr. Swanson, Understanding Scientific Research (A Handbook for the Social Professions) (1992) and Research Methods for Problem Solvers and Critical Thinkers (2009) scheduled for publication in 2011.

CONTENTS

§

APPENDICES:

Foreword

Science is concerned with consistent thinking and observation. Consistent systems of thought have been developed that make it possible for two independent persons who are provided with the same premises to arrive at the same conclusions. Likewise, methods of observation have been devised to insure that independent observers can record a particular observation in a manner that recognizes it as the same observation. Such public thought and observations processes are fundamental to all scientific investigation. Measurement theory provides a basis for consistent quantitative thought systems that describe observations.

Daily, the social professions are confronted with an increasing volume of relevant "scientific" information. because they are one step closer to the ultimate application of human knowledge-- life applications -- than applied sciences, professions constitute a final mixing bowl of disciplinal knowledge. The information they use to construct their purpose-directed thought systems is often the product of science, both pure and applied, as well as physical, biological, and social. In the past, the training of professionals has not given adequate attention to the questions of what constitutes scientific information and how scientific research relates to the professions. This text is an effort to provide introductory answers to these questions in a manner that facilitates intermediate and advanced investigations by its readers.

Complexity is a major obstacle to scientific investigation. Consequently, simplification methods must be employed. Modern scientific research widely uses statistical inference to overcome the complexity of measuring all objects of interest. Generally, statistical inference is based on a combination of measurement theory and involves procedures of hypothesis testing. These three important elements of scientific research are related and, in fact, integrated. Finally, hypothesis testing is achieved within the confines of controlled observation by proper research design. Here it is important to realize that all "scientific studies" are not necessarily good science.

This text is both an introduction and a handbook for social scientific research. This dual objective is accomplished by succinctly over-viewing important building blocks of scientific research in the text itself. The best use of this book is to first study it in the context of a philosophy of research, followed by intermittent references to it as one progresses through an in-depth study of scientific research. For those who use it in formal research degree programs, it is suggested that the initial study of the text be early in the program and that it then be used to keep the "big picture" in focus as the candidates progress.

— Professor G. A. Swanson, PhD

CHAPTER 1
INTRODUCTION
COMPLICATED VS. SOPHISTICATED

THEORY

INDUCTIVE LOGIC

DEDUCTIVE LOGIC

OBSERVATION

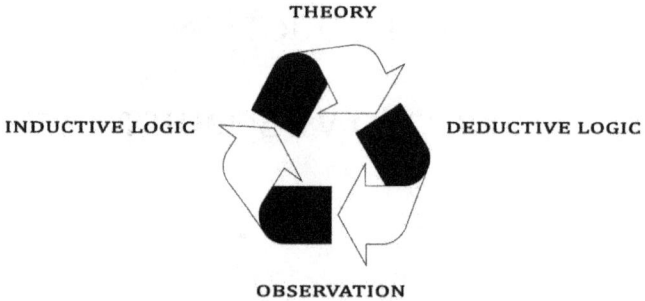

The Wheel of Research
Hunch + Assumptions + Assertions + Objectives + Lit Review +
Research Hypotheses + Statistical Hypotheses + Observations +
Conclusions + Generalizations

Research is not complicated; it is sophisticated.
Social scientific research is an invention of humans and
has been developed over time. This refinement causes
complexity and specialized knowledge to be required to
simplify the methods and the process. Once those methods
and procedures are understood the methodology no longer
appears to be complicated.

A Logical Dichotomy
A logical dichotomy in the research process is clear: first,
the process of developing a plan or proposal, and second,
the gathering and analysis of data and preparing the
conclusions and implications. The proposal is a "we –
process." The academic institution has regulations and
procedures which must be followed. A comprehensive
review of relevant literature is necessary to place the
research problem in the context of the thinking of others.
The proposal process is also guided by institutional mentors
and advisors. This makes the research proposal a "we
project." Notwithstanding, the "we-ness" of the proposal
process, the academic research process, that of gathering
and analyzing data, testing hypotheses, and reporting
findings is a most "personal process." The actual research is
a "me" thing!

Research must have a Blueprint

Although most of the terminology associated with research
design originated with experimental research, all research
must have a blueprint designed to function within a virtual
laboratory that clearly integrates the process and structure
that will assures valid conclusions. All research problems fit
into a larger context. Consideration of the area of concern
should deal first with the general context of the problem,

and then proceed logically to the specific issues. This is a kind of hour-glass effect.

The Hour-glass Effect

Beginning with the broad based problem, the critical thinking must then narrow to the specific aspect of the problem with which the research will deal. This produces a focus on the problem that leads logically to assumptions and assertions about the relationship of the problem to antecedent causes and factors that may have influenced the creation of the problem. Such a narrowing of focus leads the researcher to develop research objectives to guide a relevant literature review. Next, based on research objectives and the literature review, comes the development of a data-gathering tool and a specific plan for sampling a population, making observations, analyzing data, testing hypotheses, and arriving at conclusions and implications.

The researcher now comes to the other part of the hour-glass with critical thinking that broadens the understanding of the conclusions and implications. This will facilitate the generalization of the findings back to the population from which the sample was taken. With this the researcher has a proposal or a blueprint of the procedures to be followed to construct and guide the research process and answer the research questions and interpret the results.

A reasonable approach to social research is to learn the home keys and develop a certain muscle memory of the process. In this way, the research becomes automatic and comfortable process. When one learns the home keys, they enter a comfort zone and the process is no longer complicated.

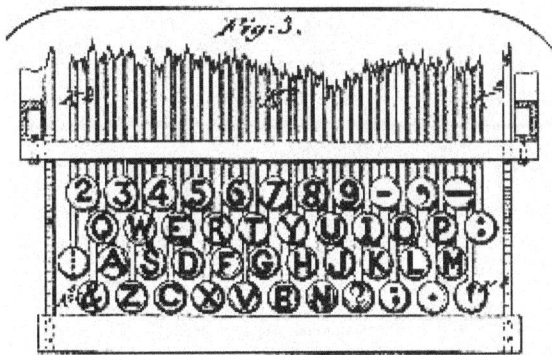

**The QWERTY keyboard shows both
Q-W-E-R-T-Y and home keys known as home row.**

CHAPTER TWO

HOME KEYS ON HOME ROW

Topics Discussed in Chapter Two
- The Home Keys On The Home Row
- Home Keys Unlock A Process
- Early Typewriters
- Only Capital Letters
- Modern Computer Keyboard
- Most Things Seem To Accumulate
- The Home Keys Are Asdfjkl(;)
- Answers To Big Questions
- History Of It Processing
- A Century Of Word Processing
- A Lapsed Dependence On Others
- Exposure To The Home Keys Came Over Time
- Don't Start Yet

THEORY

INDUCTIVE LOGIC 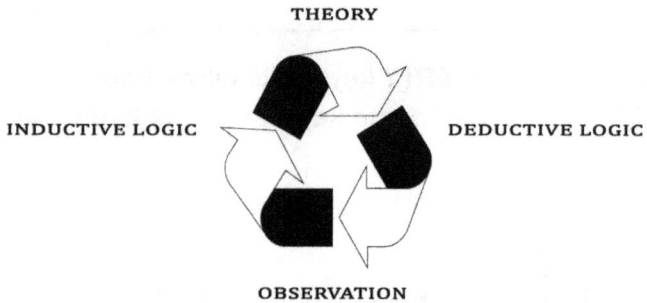 **DEDUCTIVE LOGIC**

OBSERVATION

The Wheel of Research

Hunch + Assumptions + Assertions + Objectives + Lit Review +
Research Hypotheses + Statistical Hypotheses + Observations +
Conclusions + Generalizations

T he Home Keys on the Home Row

When an operator learns the home keys on a keyboard they
enter a comfort zone. They are now ready to do productive
work. The home keys are easily identified and must be used
together to effectively produce a finished product. These
keys are a kind of cursor that marks the starting place and
tells an operator exactly where they are in the process
and points to what is to be done next. The home keys of
social research are as useful in developing and designing
a research project as were the home keys to all past and
present keyboard operations. The identified steps or keys
that unlock research development are logically obvious.
However, the unorthodox use of the keyboards on some
modern gadgets could spell an unscientific approach to
research. Those who did not learn the home keys on the
keyboard may be tempted to approach research without
learning the basic steps of the process. The function of
muscle memory is not sufficient for research.

Home Keys Unlock a Process

The home keys are familiar places that unlock a
sophisticated process. These keys operate the lock that
blocks entry and obstructs the way forward and prevents
the completion of something of value. The home keys,
once learned and understood, provide a full understanding
or explanation of the way forward or the methodology
required to produce a product. These keys are strategically
located and are vital to gaining access to and controlling the
process. They provide the answers and assist in simplifying
a complex process. The home keys become a list or outline
that can be used to construct a taxonomic identification
of the principle parts of a complex process and provide
a step by step way forward. It is clear that all persons

knowledgeable of keyboard operations understand the value of the home keys.

Early Typewriters

Many look back at the early typewriters and think they are primitive and unsophisticated; however, most keyboards in use today are called the QWERTY because of the first six letters in the top alphabet row (QWERTY). Those keys made their first appearance on a rickety, clumsy device marketed as the "Type-Writer" in 1872. Sure QWERTY appears to be complicated but it is called Universal today because of common usage. Through the years alternative keyboards have been introduced, but none seem to hit the market just right. The universal keyboard still needs the home key on home row arrangements to facilitate the process of typing or word processing. Home row is a term used by some to refer to certain keys on the center row of alphabetical letters on a typewriter or computer keyboard. Most still prefer to call them the home keys at least the people who started with the typewriter.

Only Capital Letters

The first typewriter typed only capital letters. The type-bars hung in a circle. The roller that held the paper sat over this circle, and when a key was pressed, a type-bar would swing up to hit the paper from underneath. If two type-bars were near each other in the circle, they would tend to jam when typed in succession. So, Christopher Latham Sholes figured how to take the most common letter pairs such as "TH" and make sure their type-bars hung at safe distances. He did this using a study of letter-pair frequency prepared by educator Amos Densmore. The QWERTY keyboard itself was determined by the existing mechanical linkages of the typebars inside the machine to the keys on the outside. Sholes' solution did not eliminate the problem completely, but it was greatly reduced. QWERTY's effect, by reducing

annoying jams, was to speed up typing rather than slow
it down. No contemporary account complains about the
illogical keyboard. The only major modifications of the
keyboard were a shift key to permit lower case letters and
the addition of an electronic interface.

Modern Computer Keyboard

The modern computer keyboard has added a "home" key to
return to the top of a document and memory, but the home
keys, now called the home row by some, remain central and
crucial to the use of a keyboard. A working keyboard became
standard, but without the home keys the process of typing
and today's word processing and Internet research would be
greatly hindered.

Most things seem to Accumulate

Similar to dust under the bed, word processing and research
have both evolved over time. Early the use of Boolean Logic
was common in qualitative research. The laws of Boolean
algebra (logic) can be considered self-evident when certain
equations are provable propositions that are true for
every possible assignment of 0 or 1 to a set of variables.
The axiomatic approach is complete in the sense that it
confirms respectively neither more nor less laws than the
sound and reasonable approach. As research methods
developed and moved into more quantitative areas, the
use of modern statistical procedures were initiated. It
is just these methods that create the apprehension and
anxiety about social research. Remember, social research
is not complicated; it is sophisticated. For what it is worth,
statistics are much easier than Boolean logic. Learn the
ten home keys below and the whole process will become
simplified and the process will move forward smoothly.

The Home Keys are ASDFJKL(;)

The present Home Keys are the secret to word processing. From these keys, with a little muscle memory, a person can reach all the other keys to process a document with speed and accuracy. These keys unlock the mystery of word processing. The simple keys to social research are logical steps in the process. A researcher can make all the necessary steps in order to complete a valid research project.

Key (A) - All research is geared toward the solution of a problem. The area of concern must be engaged with the interrogative pronouns (who, what, when, where, how) to see all aspects of the dilemma and begin to understand possible antecedent explanations for the problem or the possible factors that influenced the creating of the present area of concern.

Key (S) - See all aspects of the problem and delineate assumptions from the information gained through the interrogative process and initial assertions based on assumptions made about the related aspects of the problem.

Key (D) - Delineate preliminary assertions made about the relationship of the problem with other factors based on the assumptions and initiate a review of relevant literature through a library or search engine on the Internet using key words related to the problem and the possible antecedent factors. The review of relevant material will assist in developing objectives for the research. These objectives will then guide a more comprehensive review of literature to place the problem in the context of what others have written about the issues involved. This may also point out areas to be omitted from this particular project.

Key (F) - Finding relationships between key factors is the goal of a comprehensive review of relevant published works including journals. The process is based on the determined assumptions of the research. This will assist in refining the objectives where data has already been found and possible new areas where an additional objective should be determined. Once sufficient relevant data from books, journals, and other available sources are gathered, one can turn the objectives into research hypotheses. To do this one must assert that there is a relationship between the problem and some other factor.

Key (J) Justify the research process by developing a survey or interview instrument to gather data that adequately represents the research hypotheses. Sufficient questions related to each hypothesis must be included to provide enough information to adequately test the validity of the hypothesis. The most difficult part of the research process can be the development of the instrument.

Key (K) Knowing the research hypotheses, the next step is to develop the data collection tools instrument to gather data needed to test the hypotheses. The quality of research largely depends on the appropriate development and use of a data gathering instrument. A survey questionnaire designed around the proposed hypotheses is an important and time consuming aspect of valid research. One must clearly understand the difference between open-ended questions and closed questions in constructing an instrument.

Appropriate data gathering tools that cover all important variables are crucial to valid research. Normally this data will be gathered from a sample scientifically selected from the population using closed questions assessed on a five-point

Likert scale; such as, 1. Strongly agree; 2. Agree; 3. Not sure/no opinion; 4. Disagree; 5. Strongly disagree.

Key (L) Linking the research hypotheses to the construction of statistical or null hypotheses requires making a statement of "no relationship" or "no difference." Null hypotheses not only have the comparison of two factors, the problem and a possible antecedent factor, the null must also have a statistical procedure identified based on the kind of data one acquires from the instrumentation. The data must be tested to accept or reject the null. If the null is accepted, the research hypothesis is not supported or confirmed. If the null is rejected that there is no relationship or no difference, then the research hypothesis is supported or confirmed.

Key (;) Since a semicolon in writing is used to indicate a pause in the process, this research key suggests that all research projects deserve a pause or two to reflect on the data and the analysis before conclusions are made or an application of the results are generalized and the significance of the research are determined. Patience is required. Valid research cannot be rushed. Reflection is a useful tool in research.

Answers to big Questions

Now you have answers to the big questions asked earlier and you now know something that no one else on the planet knows. That is the exciting part of social research. It is worthy of the effort to learn the home keys and unlock the secrets of solving problems that concern you, your family, your work, your community, your church or faith-based organization, and perhaps your community. Take another look at the home keys as they relate to the history of IT or word processing.

History of IT Processing

The history of information processing demonstrates a natural progression and improvement. Each step was painful for the participants but productive for the workplace. At first leadership chose only men as private secretaries or assistants to process and pass on corporate and institutional information. The primary qualification was a certain trustworthiness and the ability to formulate information in cursive longhand with the strokes of the letters joined with excellent penmanship. Then came the machine assisted writing. At first the machine was called the Type Writer, later the person using the machine, usually a female, was called a "typewriter." When both men and women were hired to operate the typewriter, they were called typist. This was progress, but the typewriter was still used to process the thoughts and plans of others. There were no creative initiatives or personal involvement in the process.

A Century of Word Processing

During a century of word processing on typewriters, the written word was distinct from the author. The person transcribing the words of corporate and institutional leaders needed only to manually process the information without any creativity involvement. In other words, the word processor depended totally on another person for creation of a document. Of course, the system of shorthand required a secretary to know sentence structure and language rules. Still the word processor was not the creator of the document. Information processing was based on the intelligence and creativity of another. This began to change with the advent of computers and was accelerated by the Internet.

A Lapsed Dependence on Others

Presently it appears that Information Technology personnel have lapsed into a dependence on others to create the hardware, software, and the required procedures necessary to process information. What is needed is a method to restore creativity to information processing. This can be done by understanding the progression of social scientific research designs, the formal nature of research writing, the procedures for data gathering, and the creative ability to use statistical procedures to analyze the data and make logical and valid conclusions that will stand the scrutiny of others.

Exposure to the Home Keys came over time
Personal experience of the word processing aspects of writing came over time. In the Eighth Grade the Typing Teacher came to Study Hall and asked for volunteers for the three vacant machines in the typing class. The teacher said "Typing class is for girls." Two girls raised their hands, there was one more opening, but no one volunteered. With a strong zeal to learn, the teacher reluctantly granted me permission for one semester of typing. Sure it looked complicated because the type-writer had black keys with no letters or numbers. There was a poster of the keys on the wall and the secret was learning the "home keys." The "home keys" had indentions so they could be found without looking down. Facing the wall, typing became a simple process. Just learning the home keys and a little muscle memory developed with practice was all it took to produce an excellent document. This muscle memory related to the home keys was similar to a new student learning the piano. In fact my sister taught me finger exercises on the piano to increase my typing speed.

My high school and college work was done on a manual portable. Then electric typewriters came along and greatly improved production. Term papers in graduate school, two master's thesis, two doctoral dissertations and 45 books later, my appreciation for the home keys and one semester of typing is deep and abiding. Just as the process moved from a black keyed manual in the classroom to a portable manual typewriter, to an IBM electric, then a Mag Card electric with some memory, next the early computers, and now desktops and laptops with memory and speed that puts the old industry main frames to shame. Innovation continues!

All of this was opened to me because of an opportunity to learn the home keys. What doors of opportunity would open if you learned the home keys of social research? Each of the above steps took courage and the overcoming of some fear factor. This text is designed to remove the fears associated with research and particularly statistics. Research is not complicated; it is sophisticated. That is, when one learns the home keys, the basic rules become simplified and many doors otherwise closed are opened. This text is designed for student, IT personnel and computer literate and Internet savvy individuals interested in social research.

In sum, this text has four objectives that are pursued using systems thinking, i.e., (1) to provide the keys to understand the general philosophy of research, (2) to enhance the use of key research methods, (3) to suggest that these key methods may be used constructively as "home keys" by students and societal professionals including individuals involved in IT, computer science, and search engines, and (4) to remove the opposition and resistance of such individuals to use these social scientific methods.

Of course, it will take some cognitive readjustment and a little positive thinking, but the reward will be worth the effort. Learning the "home keys" of research methodology and understanding the statistical model of testing hypotheses using commuter software can produce a dynamic process that will change your life and career.

The same is true in social research. Early procedures required complicated Boolean algebra (logic), laboratory observations and slow and painstaking assessment of data. As alternative paths were discovered and quasi-experimental designs were introduced the shift moved from the laboratory to virtual laboratory designs and more sophisticated statistical procedures were employed. Modern and powerful statistical procedures supported by computer stat programs (WINKS, SPSS, SAS, etc.) also facilitated the social aspects of research and made it easy to test a hypothesis.

Don't Start Yet
Yes, it will be relatively easy, but don't start yet. A research project initiated prematurely can become an accident just waiting to happen. First, learn the "home keys" and note what could happen if one initiated research without the proper understanding of the problem and a workable plan. Information must be gathered in a controlled manner based on prior knowledge and analyzed properly to make a valid conclusion. Those who already have IT skills and are Internet savvy can solve problems that perplex family, work, community, established religion, or society at large by using simple keys. Using the keys is similar to using a combination lock. The process must have a specific order, use the correct numbers, and be done in reasonable time. Learn what not to do by reviewing the next episode about students trying to study a blind elephant.

STUDENTS AND A BLIND ELEPHANT

Topics Discussed in Chapter Three
- Much Is Left Unknown
- A Simulated Class Assignment Project
- A Simple Survey Design
- A Double Blind Study
- No Real Plan Or Design
- Based On Prior Knowledge
- Two Big Questions
- Class Group Project Report
- Difficulty In Describing A Blind Elephant

THEORY

INDUCTIVE LOGIC

DEDUCTIVE LOGIC

OBSERVATION

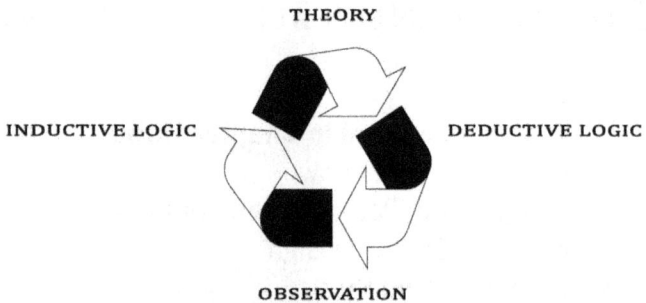

The Wheel of Research
Hunch + Assumptions + Assertions + Objectives + Lit Review +
Research Hypotheses + Statistical Hypotheses + Observations +
Conclusions + Generalizations

Much is Left Unknown

Simply put, social research is similar to a man blind from birth attempting to describe an elephant he has never seen. The researcher may learn a great deal in the process, but the picture would still be incomplete. So goes the research process. One learns many things, but in the end much is left unknown. Those items remaining unknown become the subject for the next research project.

A Simulated Class Assignment Project

There are lots of stories about the blind men who tried to describe an elephant, but have you heard about the lazy first year research students who attempted to do social research on a blind elephant? Well it is a sad case that breaks all the rules of social research. The students being ill informed on the research process initiated a research project prematurely. The outcome was predictably bad.

A Simple Survey Design

The students decided that an uncomplicated survey design with a population of "one blind elephant" would be a simple project. Rushing into their homework assignment, the students immediately began to "observe" the blind elephant and ask each other questions using the basic interrogative process, such as:

Why is the elephant blind?
How long has the elephant been blind?
Where was the elephant born?
Who were the parents?
How old is the elephant?
Why does the elephant just stand still?
Why doesn't the elephant have any friends?
Where and when does the elephant bathe?

A Double Blind Study

Since the elephant couldn't answer the questions, one student suggested that they do a "blind" study. One of the students was blind, so another suggested they do a "double blind" study. They had heard that a "double blind" study supposed to protect them from their bias against blind elephants. So they proceeded,.....or started to commence to begin doing little but thinking less. After all it was just a little old class assignment. It didn't appear to be important. Who cares about a blind old elephant anyway was the general consensus. At last, they agreed on something.

No Real Plan or Design

With no real plan or research design, with little understanding of methodology and no delegation of duties, the group went busily about the business of gathering information about the blind elephant. They had decided on a "touch and count" process to gather information so they touched and counted. As the process continued the blind student saw things one way, the short student saw different things, and the tallest student saw more that them all. Some made observations from the front, others began their visual tour from the back, still others looked first at the sides. One of the student's father was a Dentist and became enamored with the elephant's tusks. The blind student kept asking, "What color are the eyes?" A hearing impaired student was intrigued with the large ears of the elephant. One of the students was traumatized because the elephant was chained by the foot and couldn't walk feely. Yet they were learning something about a blind elephant. Perhaps the group was learning a little about research, too. If one can learn from failure?

Based on prior Knowledge

Based on prior knowledge of other animals and the human anatomy, they attempted to identify and count understood elements; such as, appendages, eyes, ears, etc. When the final written report was made based on the "touch" observations and the "double blind" study, the elephant appear to be some prehistoric animal extinct for centuries. Without the tools of social research and the control brought to the process by a research design, there was little hope of an informed outcome. Then there was only one elephant and nothing around with which to compare their observation. It was a mess.

Two Big Questions
The big question the Professor would ask, "Was the data reliable? That is, did they all get the same answers?" Obviously, they did not get the same answers. The next question, "Were your findings valid? That is, are your conclusions correct?" Without the control of a valid research design and the ability to compare the blind elephant with other elephants, the findings could not be generalized to other elephants. The homework assignment was an obvious failure. Hopefully, the students learned a good lesson from the frustrated effort to describe a blind elephant. At least, they turned in a report.

Class Group Project Report
We all participated in the observation, but we each touched and observed a different part of the elephant. This brought considerable disagreement. Here are the facts on which we agreed. The elephant was blind and had no friends. The elephant doesn't walk around; it is chained by the foot. The elephant has big front teeth and a long hose of a nose. The elephant has big feet and fat legs. The elephant has big ears and a short tail. The elephant was taller in front than it was behind. The elephant had hard skin and was in bad need of body grooming. The elephant could use a bath and a little perfume, too.

Difficulty in Describing a Blind Elephant

We did descriptive research on a blind elephant. There was difficulty in describing the elephant because there were so many different opinions. For extra credit we decided to work together, and pool our knowledge about the elephant. We have presented a composite drawing of the elephant so you can see our observations. We have showed you both sides of the elephant. We wanted you to know how complete our observations were of the blind elephant. We left the chain off the foot in the drawing, because it upset one of the students. We hope you like our report and the drawing. Since we didn't come up with an accurate count on the number of appendages, you can count them yourself from the drawing. To some the elephant appeared to have too many legs.

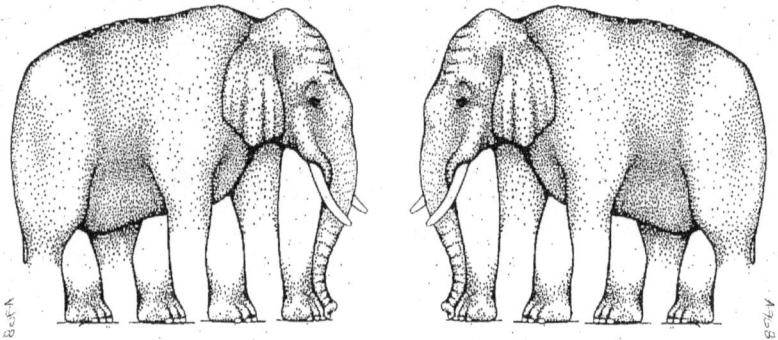

Respectfully,
Mary, Sue, Sara Jane, Beverly, Elizabeth, and Charlie

CHAPTER FOUR

PLANNED AND DETAILED PROCESS

Topics Discussed in Chapter Four
- A Detailed Process
- Format Of Research
- A Conceptual Framework
- Consistent Thinking And Observation
- A Major Obstacle To Investigation
- Background Of Scientific Research
- The Idea Of Scientific Prediction Emerged
- Choosing A Research Problem
- Preparing A Research Investigation
- Lewin's Force Field Theory
- Early Education Exposure
- Lewin Developed Principles
- Dissatisfaction Or Frustration

THEORY

INDUCTIVE LOGIC

DEDUCTIVE LOGIC

OBSERVATION

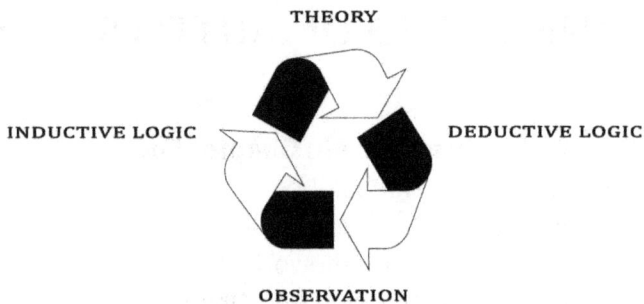

The Wheel of Research

Hunch + Assumptions + Assertions + Objectives + Lit Review +
Research Hypotheses + Statistical Hypotheses + Observations +
Conclusions + Generalizations

A Detailed Process

Research is a planned and detailed process. Hunches, expectations, and assumptions should be developed early. A plan of action should be followed by which another independent researcher using the same plan would achieve similar results. Such plans should include objectives, methods, and procedures, the significance of the research, pertinent literature review such as previous research that relates to the problem, descriptions of the results obtained by the other related research, and other supporting information. Methods of research tend to vary from discipline to discipline, but research generally follows similar basic comprehensive plans. A prospectus is basically a means of selling a research idea, either to academia or a funding source. For academia, the proposal is prepared to provide evidence that a study can be completed and replicated. For business and industry, it provides assurance that the expected benefit is worth the investment.

Format of Research

The actual format of research is normally specified by an academic program, a discipline, or a research industry. A proposal for a theoretical study employing qualitative methods that emphasize intuitive induction, description, and the study of environments, attitudes, and perceptions should contain a specific statement of approach, methodology, extensive literature review, but no testable hypotheses. Research that employs quantitative methods should require a detailed proposal, presenting a specific focus supported by a literature review and at least one testable hypothesis. Such a proposal should include a specific design for the research with descriptions of sampling procedures, instruments to be used, data-gathering procedures, and data analysis procedures.

A Conceptual Framework

This text incorporates a systems view of existence (Chapter 4). This view makes it possible to efficiently integrate diverse disciplines of thought. James Grier Miller's living systems theory is used to provide a comprehensive conceptual framework by which knowledge accumulated in the various scientific disciplines may be interrelated and studied.

Dealing with Consistency and Complexity

Consistent thinking and observation is a primary concern of science. Complexity is a major obstacle to valid and reliable research. Consequently, a master plan and the process of simplification must be used. The use of statistical inference can overcome the complexity of dealing with all subjects in a population. The statistical testing of hypotheses together with the interdisciplinary thinking of James Grier Miller's living systems (LST) may simplify the approach process.

Background of Scientific Research

The etymological meaning of the word research is "seek again." Today this word is commonly used to describe a wide range of activities concerned with human inquiry. Literally, the derived meaning suggests two fundamental characteristics: (1) diligence of inquiry and (2) a temporal relationship only to the present and past, although a basic purpose may be to predict and control the future. Research is a snapshot of a fixed time period. Perhaps it is this latter characteristic that fixed the earliest inquiries on questions of origin and existence. A diligent searching again of the past reveals that existence, moving over time, has order, patterns, and sequences. The discovery of relationships that persisted over time made it possible to seek answers to questions about the future. Predictive instruments, such as calendars, were based on observed recurring patterns of the past.

The Idea of Scientific Prediction Emerged

As human orientation moved from efforts to understand the past to efforts to inquire into and control the present and the future, the idea of scientific prediction emerged. Scientific prediction is narrowly defined. It predicts the outcome of tightly controlled rearrangements of already known structures and processes. Scientific prediction relates to time only in the sense that a scientific study or experiment is a process and all processes must occur over time. The meaning of scientific prediction is nearly the opposite of the common meaning of the term prophecy, prediction of future events that the prophet can neither control nor understand.

To investigate different aspects of human existence, various groups of scholars began to devise different schemes that all used scientific prediction as the basic method of inquiry. Eventually, sets of such methods were accepted as reliable by consensus of particular groups. Different groups were concerned with different aspects of human inquiry; and, consequently, the research methodologies that evolved differed from group to group.

Choosing a Research Problem

An evaluation of the problem or the area of concern must come early in the process. When the thinking has settled on a particular problem or rather a specific aspect of the problem, one is ready to proceed with some personal considerations. Does this problem fit my goals and the expectations of others involved? Can I control my bias in spite of my personal interest and assumptions about the problem? Does research on this problem fall within my skills, abilities and my background knowledge of the general area of concern? Has available time, money, and adequate access to a population to gather reliable data been considered? Is the scope and significance of the problem worthy of my time and will others cooperate in the conduct of this research?

Preparing a Research Investigation

After selecting a problem within the area of concern that begs for a solution, the first step is problem-analysis. Accumulate the facts that might be related to the problem. Make assumptions about the problem whether or not the facts are relevant to the problem. Identify any relationship between facts that could reveal factors that precipitated or influenced the problem. Make assertions about the problem based on possible explanations for the existence of the problem. Through a brief review of relevant literature (books and journals), determine whether these assertions relate directly to the problem. Consider relationships between explanations that could provide insight into the area of concern. Do a more comprehensive literature review to trace relationships between assumptions, assertions, and explanations.

Next, you are ready to establish basic objectives for research on the problem. When the objectives have been established and carefully examined to determine suitability through relevant literature, construct research hypotheses which are basic assertions about expected relationships between relevant factors. This means the original assumptions have been adequately vetted, and objectives and research hypotheses have been established.

Social Considerations for the Research

Could a solution of the problem advance knowledge and make a contribution to my field of study? Can the results be adequately generalized to the population studied? Will the findings be of practical value? Has the literature review placed the problem in the context of existing research? Will the research replicate another study?   If so, was the subject extended beyond previous limits? Has the subject been sufficiently delimited to assure an exhaustive study yet important enough to support further study? Will the data gathered from a sample of the population be reliable and the conclusions valid? Were the methods used sufficiently rigid and the controls adequate to cause others to trust the conclusions?

A Fresh Desire toward Logical Steps

Hopefully, this work will spark a fresh desire to take the next logical steps in the information process to utilize more effectively the tools already learned by adding a research component to the mix. In this way one may advance to a more effective leadership position in the organizations and institutions of which they are a part. It will take a little courage and a little time, but the end results will be worth the effort. Learning more about the research process and how to utilize the Internet to gather information should be a liberating process that opens many opportunities for individuals who know how to use personal computers. Learning the research methods necessary to structure research instruments and gather reliable data to arrive at valid conclusions can become a force forward equal to the dynamics propagated in Lewin's Force Field Theory.

Lewin's Force field theory was an influential development for analysis in the field of social science. It provided a framework for understanding the positive and negative factors that influenced change. It considers driving forces and restraining forces that either produce or inhibit change. The theory, developed by Kurt Levin, was a significant contribution to the fields of social science, psychology, social psychology, organizational development, process management, and change theory. Hopefully, it can be used to remove the fears of students and IT personnel about moving on to social research.

My early education exposure to Lewin's Force Field Theory made an impact on my academic, professional, and personal pursuits. My understanding that one could remove as little as 10 percent of the opposition to any change and it had the same force as putting ten times the force to move the process forward. This fact has influenced many of my personal perspectives and constructs on organizational and institutional change. This is why 35 years of my academic life was devoted to teaching research design, methodology, statistics as a language rather than a math, and the Think-Plan-Organize (TPO) formula for writing. It seems that the "math" stuff scares good students away from learning the simple process and procedures required to do valid research. The home key approach simplifies the process.

From constructed theory, Lewin developed principles and applied them to the analysis of group conflict, learning, adolescence, hatred, morale, and society. This broke down fears and misconceptions about change and determined basic constructs that many utilize in understanding social change. It was not only theory, but common sense that changed the way academics and leaders handled the organizational and institutional problems related to change. It is this fear of taking on something new or a fear of the unknown that appears to be a complicating factor as well as something to do with the "math" question that caused many to resist the change necessary to become a social researcher. Those who already know the field of IT, computers, and the use of search engines are ready to learn the "home keys" of research methods and proceed post haste.

All progress and change starts with some dissatisfaction or frustration generated by the mundane work and the fear of any change. Hopefully, this text will remove at least 10 percent of the resistance to learning social research methods that generate reliable data and basic statistical processes that produce valid conclusions from that data. Just for the faint of heart, there are several software programs that function to process statistical data for the researcher. What used to take professional academics months to process can now be accomplished by students on a Personal Computer using one of many programs easily available; such as, WINKS, SPSS, SAS, and others?

Lewin's theories together with James Grier Millier's Living Systems Theory (LST) informed many of my academic and leadership concepts and constructs. It would be appropriate for all students to study both Lewin and Miller in the context of academic research.

CHAPTER FIVE

LIVING SYSTEMS AND RESEARCH

Topics Discussed in Chapter Five

- A Myriad Of Disciplines
- Types Of Systems To Evaluate Scientific Research
- Living Systems
- The Hierarchy Of Living Systems
- Social Research And The Hierarchy Of Living Systems
- Critical Subsystems In Living Systems Theory
- Levels Of Research Of Concern To Societal Professions
- Commonly Used Systems That Concern Social Research

THEORY

INDUCTIVE LOGIC

DEDUCTIVE LOGIC

OBSERVATION

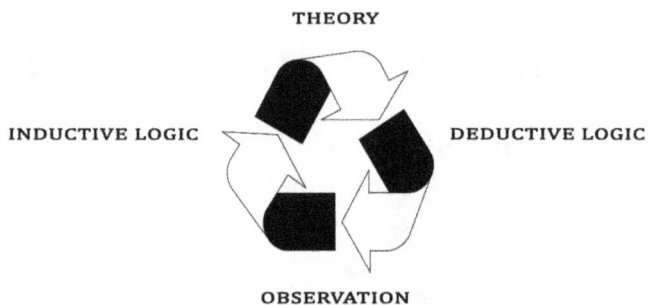

The Wheel of Research

Hunch + Assumptions + Assertions + Objectives + Lit Review +
Research Hypotheses + Statistical Hypotheses + Observations +
Conclusions + Generalizations

A Myriad of Disciplines

Today, the total effort of formal human inquiry is divided into a myriad of disciplines that investigate various systems ranging from abstract to concrete, i.e., superstitions, religions, philosophies, and sciences. Numerous disciplines exist within the sciences. Each discipline decides what methodologies are to be allowed to find its particular niche. Such division, while generally enhancing specific analytic research, produces inefficiencies in research dealing with syntheses and generally confuses informed laity and experts from other disciplines.

This situation makes it difficult for certain specialized groups, such as students, educators, religious clerics, accountants, lawyers, doctors, social workers, government ministers, and counselors, to use modern scientific research or its results to answer questions that arise in practicing their vocations. A basic objective of this text is to suggest to those individuals that scientific research methodology can be understood and used to investigate many aspects of their concerns. Additionally, the text is designed to entice them to actually use research methodology. To this end, confusion that may result from fragmenting the scientific community into disciplines is mitigated to some extent by basing the text on Miller's living systems concepts.

In particular, it is suggested that students use the terminology and concepts synthesized by James Grier Miller in his comprehensive conceptual framework termed *living systems theory* (LST). (Miller,1978). As a pedagogical tool, this framework can assist various vocations develop a general frame of reference expeditiously and systematically. New researchers may by reviewing a brief introduction to certain fundamental elements of LST. Over time, social researchers should make an in-depth study of Miller's work. However, such study should be made as a corollary to actually doing research in personal areas of employment.

Types of Systems to Evaluate Scientific Research
A useful trichotomy of the universe of all systems is as follows; (1) concrete systems, (2) abstracted systems, and (3) conceptual systems. The latter two types may be sub-typed. Although LST appears complicated, it is in reality a simplifying process to assist the overall understanding of a specific area of concern. So, in reality, LST is a system to assist thinking about problems. (Swanson/Green, 1992)

A **concrete** system is a nonrandom accumulation of matter, energy, and information in a region of physical space-time, which is organized into interacting, interrelated sub-systems or components. Examples of concrete systems are iron ore, a building, the human body, a radio, and an organization.

An **abstracted** system is a limited set of relationships abstracted or selected by an observer. Abstracted systems are studied by conceptual systems. Examples of abstracted systems are a person's mental image of a lover, a modern theologian's opinion of Luther's systematic theology, and the relationships such as length and angle existing on an object being measured and instrumentally observed. Abstracted systems may be sub-typed as *transformable* (measured) and non-*transformable* (surrogated).

A **conceptual** *system* is a set of words, symbols, or numbers, including those in computer simulations and programs, that have one or more, similarly ordered subsets. Conceptual systems have two sub-types, *quantifiable* and *non-quantifiable*.

All societal vocations are directly and obviously concerned with three levels of living systems, i.e., organism, group, and organization. Less directly, but quite obviously, they are also concerned about the levels of community, society, and supranational systems. Therefore, by the nature of their concerns, the social vocations are cross-level disciplines. LST provides a handy set of vocabulary, concepts, and postulates in a logical framework to facilitate research by such disciplines.

Living Systems

Concrete systems may be divided into two categories, i.e. (1) living systems, and (2) non-living systems. While some non-living systems may be open systems, all living systems are open systems. An open system is one with permeable boundaries. Closed systems, having impermeable boundaries, are subject to the law of thermodynamic degradation (moving toward random disorder). The definition of system (being related, interacting elements) presupposes order or arrangement. Movement toward random disorder is, therefore, movement toward system dissipation. Living systems, being open systems, overcome this propensity by importing matter, energy, and information of a higher complexity and breaking it down to repair dissipation within the system, thereby maintaining a dynamic steady state of arrangement and order over and indefinite period.

Information, in concrete systems, basically refers to the patterned arrangement of the matter and energy of the system. Its transmission on channels in space, or retention overtime, requires that the information be carried on bundles of matter-energy called markers. Information can be borne on many sorts of markers including stone tablets, papyrus, paper, neural impulses, gestures, sound waves, radio and television waves, and electronic currents. The movement of information markers though physical channels and nets constitutes communication or information flow. If the transmitter and receiver of communications use the same language or code, and if the transmission alters the behavior of the receiver in some way, the impact of the transmission is called its meaning. Money is a special information flow that emerges at the society level of living systems. In advanced economies, the monetary flow is a fundamental information flow.

Non-living inclusion e.g., buildings, alphabets, libraries, engines, and computers, in living systems are termed artifacts. An artifact may or may not be a prosthesis. Most people are familiar with the common usage of the term prosthesis to describe a device such as an artificial arm. In systems above the organism level, the term is used to describe an artifact that carries out some process essential to a living system, e.g., the accounting system, or the education system of a church or community.

The Hierarchy of Living Systems
Living systems exist in a hierarchy of complexity from the cell to the supranational system. The hierarchy, when clearly understood, assists researchers in determining exactly where to secure data to study a particular level in the system.

The study of cells, organs, and organisms remains the domain of laboratory research.

1. **Cells** - made of atoms, molecules, and organelles
2. **Organs** - made of cells
3. **Organisms** - composed of organs

Groups, organizations, and communities are the domain of social research.

4. **Groups** - made of organisms (data is gathered from individuals in a group)
5. **Organizations** - have groups as components. (Data is gathered from individuals, but the analysis of data places each individual answer together with others in the same group)
6. **Communities** - have organizations as components. (to study a community one must gather data from all the organizations that make up the community)

A study of society or supranational system is normally beyond the skill of a social researcher.

7. **Societies** - have organizations and communities as parts
8. **Supranational systems** - complexes of societies

All systems are systems within a given system, because the lower-level systems of the hierarchy are contained in the higher-level systems. The system above a given system may be termed its *suprasystem*. When one is conducting research on a given level of the hierarchy, **data from the previous level is used to study that level**. For example, Groups are made up of individuals (organisms), in order to adequately study the group one must gather data from all the individuals that make up the group. Likewise, an organization is made up of groups and one must gather data from all the groups in the organization to adequately study that organization. Of course, that data will come from individuals within the groups, but in analyzing the data the answers from these individuals must be gathered into groups and a decision based on a composite of all answers used to determine the attitude, knowledge, or behavior of the groups that make up the organization. In the early days of pre-computer assisted analysis, one had to pile up IBM punched-cards in stacks based on the individuals in the groups. Now, or course, the PC programs can do that work for the researcher. Simply "ask" the data, "How did the members of the Choir answer questions 6-11, etc." Now one understands, not the gripes or complaints of a few, but the attitude, knowledge, and/or behavior of the group (Choir members) on the issue and how that data impacts the organization as a whole.

Social Research and the Hierarchy of Living Systems
Living systems theory (LST) is a conceptual framework that integrates scientific facts about the hierarchy of organizations, communities, societies, and supranational systems. This inter-disciplinary field recognizes that these living systems differ in numerous obvious ways. Nevertheless, they are similar in many important ways. Consequently, it is useful to attempt to construct a general theory that recognizes the similarities in physical, biological, and social systems.

It is important to recognize that LST is not merely a system of analogies. Rather, because these eight levels of life share a common developmental process, as expressed in the principle of fray-out, it is possible to make cross-level comparisons by constructing cross-level experiments and research designs that produce cross-level formal identities, e.g., between organisms and groups or organizations and communities.

In this manner, it is possible to integrate much of the scientific knowledge obtained by the social, biological, and physical sciences that relates to the structure and processes at any of the levels. Anthropology, biochemistry, economics, genetics, medicine, pharmacology, physiology, political science, psychology, and sociology are almost entirely relevant while many aspects of physical science and engineering are also relevant. Logic, mathematics, and statistics provide methods, models, and simulations, including the new approaches of cybernetics and information theory.

Thus, LST may be used as a guide for integrating general scientific facts and methods into areas of applied science such as the arts of propagating religion, counseling, teaching, and accounting. While it is a general theory, the detail of its construction is such that it directly aids in constructing empirically testable hypotheses. For example, LST identifies twenty critical functional subsystems. These subsystems are part of every type of system from cell to supranational system, for instance, the input transducer is the sensory subsystem that brings markers bearing information into the system, changing them to other matter-energy forms suitable for transmission within the system.

Critical Subsystems in Living Systems Theory

SUBSYSTEMS WHICH PROCESS BOTH MATTER/ENERGY AND INFORMATION

1. **Reproducer**, the subsystem which is capable of giving rise to other systems similar to itself. The reproducer is the only subsystem that does not automatically function.

2. **Boundary**, the subsystem at the perimeter of a system that holds together the components which make up the system, protects them from environmental stresses, and excludes or permits entry in various sorts of matter/energy and information.

SUBSYSTEMS WHICH PROCESS MATTER/ENERGY

3. **Ingester**, the subsystem which brings matter/energy across the system boundary from the environment.

4. **Distributor**, the subsystem which carries inputs from outside the system or outputs from its subsystem around the system to each component.

5. **Converter**, the subsystem which changes certain inputs to the system into forms more useful for special purposes of the particular systems.

6. **Producer**, the subsystem which forms stable associations that endures for significant periods among matter-energy inputs to the system or outputs from its converter. The materials, synthesized being for growth, damage repair, or replacement of components of the system, or for providing energy for moving or constituting the system's outputs of products or information markers to its suprasystem.

7. **Matter/Energy Storage**, the subsystem which retains in the system, for different periods of time, deposits of various sorts of matter/energy.

8. **Extruder**, the subsystem which transmits matter/energy out of the system in the form of products or wastes.

9. **Motor**, the subsystem which moves the system or parts of it in relation to part or all of its environment or moves components of its environment in relation to each other.

10. **Supporter**, the subsystem which maintains the proper spatial relationships among components of the system, so that the can interact without weighing each otherdown or crowding each other.

SUBSYSTEMS WHICH PROCESS INFORMATION

11. **Input transducer**, the sensory subsystem which rings markers bearing information into the system, changing them to other matter/energy forms suitable for transmission within it.

12. **Internal transducer**, the sensory subsystem which receives, from which subsystems or components within the system,

markers bearing information about significant alterations in those subsystems or components, changing them to other matter/ energy forms of a sort which can be transmitted within it.

13. **Channel and net**, the subsystem composed of a single route in physical space, or multiple interconnected routes, by which markers bearing information are transmitted to parts of the system.

14. **Timer, the clock**, set by information from the input transducer about states of the environment that uses information about processing in time and transmits to the decider signals that facilitate coordination of the system's processes in time.

15. **Decoder**, the subsystem which alters the code of information input to it thought the input transducer or internal transducer into a "private" code that can be used internally by the system.

17. **Memory**, the subsystem which carries out the second stage of the learning process, storing various sorts of information in the system for different periods of time.

18. **Decider**, the executive subsystem which receives all information inputs from all other subsystems and transmit to them information outputs that control the entire system.

19. **Encoder**, the subsystem which alters the code of information input to it from other information processing subsystems, from a "private" code used internally by the system into a "public" code which can be interpreted by other systems in its environment.

20. **Output transducer**, the subsystem which puts out markers bearing information from the system, changing markers within the system into other matte/-energy forms which can be transmitted over channels in the system's environment

---

Source: G.A. Swanson and James Grier Miller, Measurement and interpretation in Accounting: A Living Systems Theory Approach. New York: Qurum Books, 1989. (pp. 56-57).

While the structure of a system is the spatial location
of the matter-energy in such subsystems at a moment,
the emphasis of the theory is on process. All of these
subsystems involve processes. They are dynamic. They
change over time. In fact, structure is constantly changed
over time by process. These two qualities taken together
constitute the state of a system. All living systems maintain
steady-states over time; i.e., the interrelationship of their
variables and their relationships to environmental variables
are maintained within relatively narrow ranges.

Due to its detailed constructs and relationships, LST
identifies a host of relationships that may be measured
to assist with an understand of living systems concerning
societal vocations, e.g., structural relationships —
containment, number order, positions, direction, size
pattern, and density; process relationships -- containment
in time, number in time, order in time, position in time,
direction in time, duration and pattern in time; spatio-
temporal relationships -- action, communication, direction
of action, pattern of action, and entering or leaving
containment.

Levels of Research of concern to Societal Professions
While all of physical existence (formed of concrete systems)
is marvelously unified and its components interdependent,
those engaged in the societal vocations are immediately
concerned with organisms, groups, organizations,
communities, and societies.. These terms are in common
use among most societal vocations, although sometimes
used loosely. Those engaged in Faith-based operations
typically term these levels the individual, the family, the
place of worship, and the community. At these levels, it
is convenient to view concrete systems in terms not only
of matter and energy, but also certain common types and
combinations of the two; five fundamental concrete flows

constitute the objects of scientific investigation at these levels, i.e., materials, energy, personnel, communications, and money.

Scientific research is concerned with the levels of organism, group, organization, and society. When one is studying a group, the data comes from the organism level; studying an organization, the data comes from the group level, etc. The area of religion and faith-based issues are left to the methodology used by theologians and faith-based scholars, except as the study relates to attitude, knowledge and behavior of individuals.

Modern science has become compartmentalized based on conceptual issues. Separate disciplines are emphasized by organizing universities into departments. Academics are rewarded for becoming expert in a specialty or sub-specialty. Although the major work of science must be done by specialists, it is important that the work of specialists contribute to a mosaic that advances the general knowledge of human existence.

Most of the major problems today can be solved only by multi-disciplinary approaches. For example, the effect of natural calamities, epidemics, and nuclear disasters can be studied only by some combination of disciplines such as economics, psychology, biology, medicine, atmospheric sciences, nuclear physics, engineering, accounting, and law. Several of these disciplines involve social professionals, as well as scientists. LST provides a theoretic framework to integrate the work of such specialties. Consequently, it may assist the understanding of the various disciplines and properly apply their ideas and research methods to particular problems.

Commonly Used Systems that Concern Social Research

Basic Institutions are related to the Individual. A Higher Power reaches the individual in the context of basic institutions. All institutions must work within these boundaries to be effective.

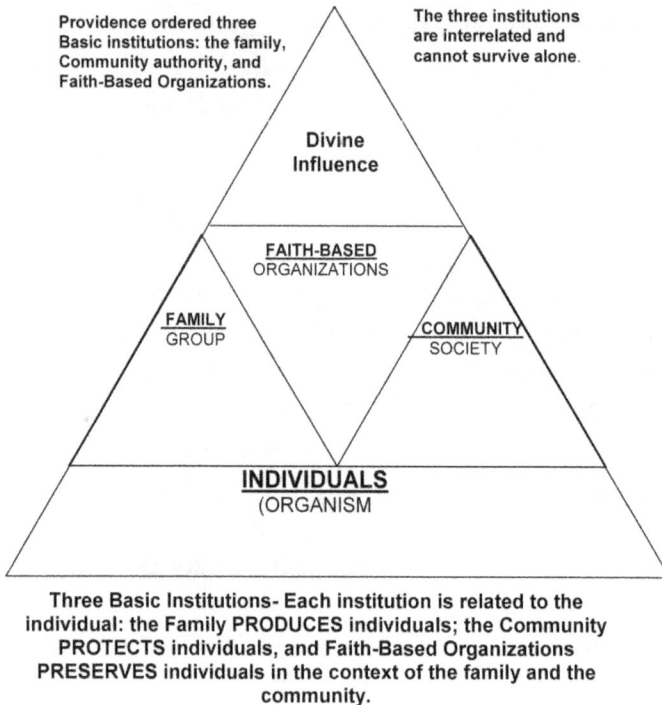

Providence ordered three Basic institutions: the family, Community authority, and Faith-Based Organizations.

The three institutions are interrelated and cannot survive alone.

Divine Influence

FAITH-BASED ORGANIZATIONS

FAMILY GROUP

COMMUNITY SOCIETY

INDIVIDUALS (ORGANISM

Three Basic Institutions- Each institution is related to the individual: the Family PRODUCES individuals; the Community PROTECTS individuals, and Faith-Based Organizations PRESERVES individuals in the context of the family and the community.

Good standard texts are available that deal adequately with the sophisticated aspects of research methodology and design. This book addresses scientific research in the context of its usefulness to individuals practicing in people-oriented organizations such as Faith-based groups, service organizations, and small governmental units. The book may be entry level learning for some, a review of basics for others, or a new way of thinking about research for still others. In any case, it presents a coherent philosophy of social research at a level most informed individuals can understand.

The body of knowledge discussed incorporates advanced statistical procedures. The discussion, however, concerns why the procedures are used. Any discussion of how a procedure works is incidental to explaining why it is used. Detailed discussion is used only to illustrate the rigor of the logic involved in modern research methods, not to teach any statistical operations. References to books and journal articles are provided for readers who desire to study the various procedures that are mentioned.

A general philosophy of research is developed in sufficient detail for applying it to specific research projects. This philosophy ties many operations together so that the individual may understand what is meant by social scientific research as opposed to other forms of human inquiry, what operations it permits, and why it limits the set of permitted operations. How to execute specific operations is not discussed in sufficient detail for this text to stand alone as a detailed guide for a specific research project. However, we do provide references to in-depth guides for specific operations, which themselves serves as starting points for studying related issues.

An inexperienced researcher must study in detail the specific methods to be used to investigate particular questions. This text provides a framework that prevents "losing sight of the forest for the trees." Certain fundamentals are needed to understand social scientific research. Most individuals are bombarded with research related to society and their vocations. These individuals need to understand the social research process to evaluate and interpret current research that impacts on their particular areas of interest.

All professionals should develop the ability to distinguish between good and bad research done by others. To do this, one needs a general understanding of the procedures and logic that make research acceptable. To determine the importance and implications of research findings, one needs to be familiar with current research methodology. Because research is method centered, the best way to learn it is to do it. Consequently, IT practitioners and students are encouraged to engage in research.

CHAPTER SIX

DESIGN AND VALIDITY

Topics Discussed in Chapter Six

- The Research Problem
- Features of Valid Research Designs
- Nine Basic Methods of Research
- A Purpose/Problem Statement
- Constructing Testable Hypotheses

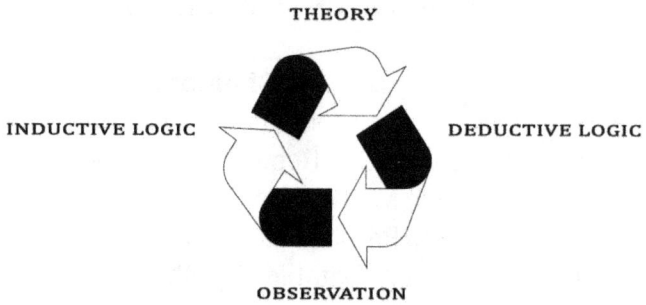

THEORY

INDUCTIVE LOGIC   DEDUCTIVE LOGIC

OBSERVATION

The Wheel of Research

Hunch + Assumptions + Assertions + Objectives + Lit Review +
Research Hypotheses + Statistical Hypotheses + Observations +
Conclusions + Generalizations

The Research Problem

After the research problem is identified, a purpose or problem statement must be developed to indicate precisely which aspects of the problem will be examined. Most social problems are too complicated to consider as a whole; therefore, the background of the problem may be an hour-glass approach that identifies the general problem and the various aspects of the problem that are visible in society. Notwithstanding, the problem identification must culminate in a description of the portion of the problem the research will observe. The observation strategy must be articulated to include assertions of either theoretical contributions or practical solutions related to the specific problem. Once the structure and strategy for this process is clearly delineated, the researcher may determine the contribution to scientific understanding the research will make. This prepares the researcher to support a significance statement related to the research outcomes. Such a statement explains what contribution the findings make to the field of knowledge.

Features of Valid Research Designs

Specific objectives are considered to be measurable if two or more people can easily agree on all the words and terms used to describe the purpose of the research. Sound research design must provide both control and comparison. The term design in the context of research refers to the way the researcher arranges the environment in which the research takes place. Research must occur in an arranged or designed environment. The design is an essential component of an environmental arrangement in which data are collected, analyzed, and interpreted.

A sound choice must be made of a population affected by the problem so that an adequate scientifically selected sample may be selected as a subset of a population. To achieve a representative sample, the researcher must use an unbiased method to determine participants and obtain an adequate number based on the size of the population to which findings are to be generalized. At times concurrent controls without random assignment are required because of the nature of the population. There is an old example of such in the Book of Daniel (1:1-11). There was no way to randomly assign the "children" to groups between those who ate the Kings mean and wine and the "children" who ate (pulse) vegetables and water. Quasi-experimental designs permit a variety of sampling and/or the selection of participants. As long as all in each group have sameness the comparability is reliable. Another example of research from the ancient holy writings is Luke's Gospel. Luke uses words found only in Greek Classics; such as, Hippocrates. It is formal and academic. 1. Many have undertaken 2. It seemed good to me 3. To compile a narrative, 4. To write an orderly account, 5. As they were delivered to us, 6. That you may know. (Luke.1: 1-4)

Reliable and valid instruments – a reliable instrument is one that gets consistent results; a valid instrument obtains accurate results.

Appropriate analysis of data – conventional statistical and other scholarly methods may be used to analyze findings. The choice of method depends on the size of the sample, the kind of data gathered, and whether the purpose of the research is description, comparison, association (correlation), or prediction.

Accurate reporting of results – to report results fairly and accurately, the researcher must stay within the boundaries set by the design, sampling methods, data collection quality, and choice of analysis. A knowledge regarding the ways to use tables and figures improves the presentation of data.

Findings must be replicable – the transparency of methods and procedures enables other researchers in the field to test the findings, or results of the research. When findings are replicated, the original research is judged more reliable and valid.   This increases the circulation of findings in the body of literature.

Nine Basic Methods of Research

Historical Method - Reconstructing the past objectively and accurately, often in relation to the tenability of a hypothesis.

Example: A study reconstructing practices in the teaching of spelling in the United States during the past fifty years; tracing the history of civil rights in the United States education since the civil war; testing the hypothesis that Francis Bacon is the real author of the "works of William Shakespeare."

Descriptive Study -To describe systematically a situation or area of interest factually and accurately.

Example: Population census studies, public opinion surveys, fact-finding surveys, status studies, task analysis studies, questionnaire and interview studies, observation studies, job descriptions, surveys of the literature, documentary analyses, anecdotal records, critical incident reports, test score analyses, an normative data.

Developmental Study - To investigate patterns and sequences of growth and or change a function of time.

Example: A longitudinal growth study following an initial sample of 200 children from six months of age to adulthood; a cross-sectional growth study investigating changing patterns of intelligence by sampling groups of children at ten different age levels; a trend study projecting the future growth and educational needs of a community from past trends and recent building estimates.

Case and Field Study - To study intensively the background, current status, and environmental interactions of a given social unit: an individual, group, institution, or community.

Example: The case history of a child with an above average IQ but with severe learning disabilities; an intensive study of a group of teenage youngsters on probation for drug abuse; an intensive study of a typical suburban community in the Midwest in terms of its socio-economic characteristics.

Correlation Research - To investigate the extent to which variations in one factor correspond with variations in one or more other factors based on correlation coefficients.

Example: To investigate relationships between reading achievement scores and one or more other variables of interest; a factor-analytic study of several intelligence tests; a study to predict success in college based on inter-correlation patterns between college grades and selected high school variables.

Casual Comparative or Ex-Pos Facto Research - To investigate possible cause-and-effect relationships by observing some existing consequence and search back through the data for plausible casual factors.

Example: To identify factors related to the "drop-out" problem in a particular high school using data from records over the past ten years; to investigate similarities and differences between such groups as smokers and nonsmokers, readers and nonreaders, or delinquents and non-delinquents, using data on file.

True Experimental Research - Investigates a possible cause-and-effect relationship by exposing one or more experimental groups to one or more treatment conditions and comparing the results to one or more control groups not receiving the treatment (random assignment being essential).

Example: To investigate the effectiveness of three methods of teaching reading to first grade children using random assignments of children and teachers to groups and methods; to investigate the effects of a specific tranquilizing drug on the learning behavior of boys identified as "hyperactive" using random assignment to groups receiving three different levels of the drug and two control groups with and without a placebo, respectively.

Quasi-experimental Research - To approximate the conditions of the true experiment in a setting which does not allow the control and or manipulation of all relevant variables? The researcher must clearly understand compromises exist in the internal and external validity of his design and proceed within these limitations.

Example: Most so-called field experiments, operational research, and even the more sophisticated forms of action research which attempt to get at causal factors in real life settings where only partial control is possible: e g. an investigation of the effectiveness of any method or treatment condition where random assignment of subjects to methods or conditions is not possible.

Action Research - To develop new skills or new approaches and to solve problems with direct application to the class-room or other applied setting.

Example: An in service training program to help teachers develop new skills in facilitating class discussions; to experiment with new approaches to teaching reading to bilingual children; to develop more effective counseling techniques for underachievers.

A Purpose and/or Problem Statement
Just as in any written document, an early step is to determine the specific purpose or thesis for the composition. The thesis or purpose statement has three elements: **What, why**, and **who**. What are you writing about? Why are you writing (inform, persuade, or interpret)? And to whom are you writing. In the case of social research data is gathered from or about people or problems to describe, compare, or predict/explain their attitudes, knowledge, and/or behavior. Social research that culminates in a published document must have a working title, a target population, and a central guiding reason or a **"why"**. In the case of social research data is gathered from or about people or problems to describe, compare, or predict/explain their attitudes, knowledge, and/or behavior. Social research that culminates in a thesis or dissertation must have a working

title (what), a target population (who), and a central guiding reason to describe, compare, or predict/explain (why). In statistically supported research the hypothesis will have a measurement to test the relationship (**how**).

Constructing Testable Hypotheses
A hypothesis is not constructed until the theoretical construction is completed. First, you will have a **"hunch"** about the relationship of the research problem to the possible antecedent influences that created the present experience. The next step requires the use of the **interrogatives:** who, what, why, how, when, etc. This will provide some **assumptions** about preconditions that influenced the creation of the present problem.

Now you must search the existing literature and see what other research has contributed to an understanding of the problem. From this you can make **assertions.** After you are able to make relationship assertions you can determine an **objective** for your research. For each objective you must construct a **hypothesis.** The Literature Review based on your objectives/hypotheses will assist with the scope and limitations of the research and will assist with **operational procedures**. For each construct in a hypothesis you must determine an operational definition.

A testable hypothesis must include these elements:

WHO
- Identity of group(s) that are being measured (or treated) and compared. The "who" may be:
- [a] different subjects ("between groups"),
- [b] the same subjects measured at different times often called "repeated measures").
- Each hypothesis must have one, and only one, comparison.

WHAT

- State the relationship of the groups or measures.

- A *research hypothesis* predicts a difference and, preferably, its direction. A *null hypothesis* states that there will be no difference.

HOW

- Give the statistical procedure (measurement) used to test the relationship.

An Important Step

Once the methodology is determined a final step is to make a search in existing literature for material relative to the objectives/hypotheses of the research. With the basic decisions made, the logical step is to make an informed search of Journals for the past three years and published books for about six years. Finally, search for classics in the field that were quoted in either books or journals. Always search the knowledge index for dissertations similar to the current research subject. Any dissertation published on a relevant subject must be reviewed carefully. Also, review the relevant data in the classics to become adequately informed to proceed with the research project. This will prevent "reinventing the wheel" aspect of research.

LITERATURE AND RELEVANCE

Topics Discussed in Chapter Seven

- Only A Few Writers
- Key To Understanding A Subject
- Basic Research For Writing
- A Means To An End
- Gathering Data
- Before Starting A Review Of Literature?
- Questions To Be Considered
- Enlarging A Writer's Knowledge
- Reviews Should Comprise Essential Elements
- Understanding The Subject
- A Review Surveys Scholarly Articles
- Data Worthy Of Inclusion
- An Overview Of The Subject
- Similar To Qualitative Research
- Every Researchable Idea
- Keep In Mind The Purpose
- A Narrowing Of Focus
- A Review Could Be Endless

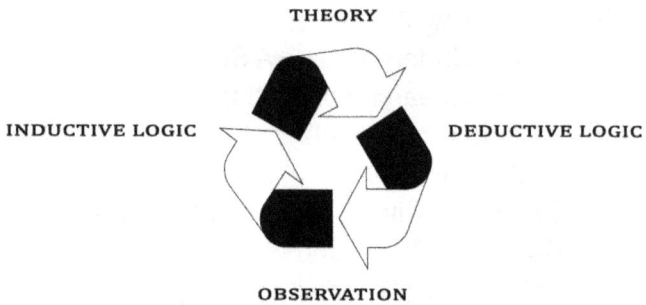

THEORY

INDUCTIVE LOGIC

DEDUCTIVE LOGIC

OBSERVATION

The Wheel of Research

Hunch + Assumptions + Assertions + Objectives + Lit Review +
Research Hypotheses + Statistical Hypotheses + Observations +
Conclusions + Generalizations

O nly a few writers

Only a few writers have sufficient knowledge to write an essay, journal article, or a book without further research. Certainly one would not attempt to write a thesis or a dissertation without an adequate or comprehensive literature review. Basic to this research is a review of relevant literature on the subject of concern. For new writers, the literature review is the least valued part of a new composition. These questions should be considered as a writer develops a literature review: 1) What is already known about the subject? 2) What are the gaps in the knowledge of the subject? 3) What areas of further study have been identified by other authors? 4) Who are the significant authors in this area? 5) Is there consensus among the primary authors on the topic? 6) What aspects of the subject have generated debate? 7) What methods or problems were identified by others studying this field that could impact the new composition? 8) What is the most productive methodology for the review? 9) What is the current status of research on the subject? 10) What sources of information or data were identified that might be useful in the new composition?

Key to Understanding a Subject

The key to understanding published writers in a given field or on a specific subject is a review of relevant material. An author should not search all the material published, but review and evaluate the relevant literature according to the research objectives guiding the composition. Data should be organized around the objectives of the new research.

A review of relevant literature does not describe one piece of literature after another, but should summarize each relevant item and do some critical assessment of material. Use an introduction and conclusion to cover the scope of the review and to formulate the question, problem, or concept for the material chosen to support new composition by grouping around the objectives of new research.

Basic Research for Writing

Authors in preparation to write an essay or book should develop an annotated bibliography as preparation for a full review of literature on the subject. The review should be a record of what has been published on a subject by respected scholars. In developing the literature review, the purpose is to determine additive and variant material that will assist in constructing the framework of the current research.

A literature review may constitute a basic research on the subject, the development of an essential chapter in a thesis or dissertation, or may be a self-contained review of a subject. Regardless the purpose for a review it should: 1) Place each author's work in the context of its contribution to the understanding of the subject under review. 2) Compare the relationship of each work to the others under consideration. 3) Identify ways to interpret or shed light on significant differences between other works. 4) Determine conflicts or data that contradicts other studies. 5) Learn areas of prior scholarship to avoid duplication of effort. 6) Decide what aspects of the problem have and have not been covered in order to determine the direction of the new research. 7) Place the new research in the context of existing literature.

Following a literature review, the author will know the ideas that have been established on a subject and be able to determine an approach to the new composition that can utilize the data already in print. The review should be guided by the purpose of the composition, the problem or issues to be discussed. A descriptive list of material available or book summaries is not sufficient, specific objectives must guide the review that will facilitate the gathering of useable material to support the new composition.

Not to be confused with a book review, a literature review surveys scholarly articles, books and other sources relevant to a particular issue, subject, or theory, providing a description, summary, and critical evaluation of each author's work. The purpose is to clearly understand the relevant literature already published on the subject. Now the reviewer knows what others have said and not said and can formulate an original approach to a research project based on new knowledge.

A Means to an End
A review of literature is a means to an end. Although the review is similar to primary research, it does not present new primary scholarship; this is left to the new composition. An adequate review of relevant literature requires four steps: 1) **Problem formulation**—which topic or field is being examined and what are the relevant issues? 2) **Literature search**—finding materials relevant to the subject being explored determines the scope. 3) **Data evaluation**—determining which literature makes a significant contribution to the understanding of the subject. 4) **Analysis and interpretation**—discussing the findings and conclusions of pertinent literature in terms of the objectives of the new research.

Gathering Data
Gathering data is an essential part of research. When a review of relevant literature permits an author to summarize and explain what others have published, the contrasting perspectives of the new composition may be clearly presented. The review will permit the writer to point out any connections between the sources and identify where

one source is built upon a prior work. This permits the new research to add to the reader's knowledge of the subject.

Before Starting a Review of Literature?
Before beginning a review of relevant literature ask these questions: 1) What specific thesis, problem, or research question does the research intend to answer? 2) What type of literature review does this research require? 3) What is the scope of the literature review? What types of publications will be used: journals, books, other document, and media data. 4) What discipline or academic field will be covered? 5) How good is the information seeking plan? 6) Has the scope of the search been sufficient to ensure all the relevant material? 7) Has it been narrow enough to exclude irrelevant material? Is the number of sources used appropriate for the length of my composition? 8) Has the literature reviewed been critically analyzed? Was a set of concepts and questions used to compare, comparing items instead of just listing and summarizing items? Was each item assessed, discussing strengths and weaknesses? 9) Were studies contrary to my perspective cited and discussed? 10) Will the literature review be relevant, appropriate, and useful in developing the new composition?

Questions to be Considered
As one reviews relevant literature certain questions should be considered about each book or article reviewed: 1) Does the author formulated a problem or issue relevant to the new composition? 2) Is it clearly defined? Is its scope, severity, relevance clearly established? 3) Could the problem have been approached more effectively from another perspective? 4) What is the author's perspective: interpretive, critical, or a combination? 5) What is the author's theoretical position: psychological or developmental? 6) What is the relationship between the theoretical and the research perspectives? 7) Is there

evidence that the author being reviewed evaluated the literature relevant to the problem or issue? 8) Does the author include literature with variant positions? 9) In an academic study, consider the value of basic components of the study design: population, intervention, and outcome? How accurate and valid are the measurements? Is the analysis of the data accurate and relevant to the research question? Are the conclusions validly based upon the data and analysis? 10) In material written for a popular readership, does the author use appeals to emotion, one-sided examples, or rhetorically-charged language and tone? Is there an objective basis to the reasoning, or is the author merely "proving" what he or she already believes? 11) In what ways does this book or article contribute to an understanding of the problem under study, and in what ways is it useful for the new composition? 12) What are the strengths and limitations of the material?

Enlarging a Writer's Knowledge
Besides enlarging a writer's knowledge about the subject of the new composition, developing a literature review permits the writer to gain and demonstrate skills in two areas: 1) information seeking: the ability to scan the literature efficiently, using manual or computerized methods, to identify a set of useful articles and books, 2) critical appraisal: the ability to apply principles of analysis to identify unbiased and valid studies.

A literature review must do four things: 1) Be organized around and related directly to the objectives of the proposed new composition, 2) Synthesize results into a summary of what is and is not known, 3) Identify areas of controversy in the literature, and 4) Formulate questions that need further development in the new composition.

Reviews should Comprise Essential Elements

Literature reviews should comprise the essential elements:
1) An overview of the subject, issues or theory under
consideration, along with new data organized around the
objectives of the new composition. 2) Division of literature
under review into categories of additive and variant material.
3) Explanation of how each concept, construct, theory, or
principle relates to the new composition. 4) Conclusions as
to the data best considered in support of the objectives of
the new composition and a decision as to how the variant
material could be presented.

Understanding the Subject

In assessing the work of each published writer,
consideration should be given to: 1) Provenance— What
are the credentials of the writer? Are arguments supported
by evidence? 2) Objectivity— Does the writer's perspective
cause disadvantage or harm to somebody or some cause?
Is variant data considered or is certain pertinent information
ignored to support the writer's point? 3) Persuasiveness—
Which of the writer's concepts, constructs, or issues are
most/least convincing? and 4) Value— Are the writer's
arguments and conclusions convincing? Does the data
contribute significantly to an understanding of the subject of
the new composition?

A Review Surveys Scholarly Articles

Such a review surveys scholarly articles, books and other
sources relevant to a particular issue, area of research,
or theory, providing a description, summary, and critical
evaluation of each work. The purpose is to offer an overview
of significant literature published on a topic and to assess
its worth and make a judgment as to how the data could be
used in an additional composition. The review of literature
should not be confused with a book review. The review of
a book deals with only one author while a literature review
deals with all the significant writers in a given field.

Data Worthy of Inclusion
A literature review is recognition of what has been published on a subject by respected authors. In developing the review, the purpose is to determine data worthy of inclusion in the new composition. Identify what knowledge and ideas have been gleaned on a subject, and what are the strengths and weaknesses of the sources. As preparation for writing, the literature review must have a guiding concept, clear objectives, and specific goals to be discussed. It is not just a descriptive list of the material available, or a set of summaries; it should glean useable material that can support a new composition.

An Overview of the Subject
Literature reviews should comprise an overview of the subject, issue or theory under consideration, along with the objectives of the literature review. Works under review should be divided into categories, those in support of and those against a particular position, and those offering alternatives with explanation of how each work is similar to and how it varies from the others. The reviewer must make decisions as to which data is best for the new composition and would make the greatest contribution to the understanding and development of the new composition.

Similar to Qualitative Research
Similar to qualitative research, development of the literature review constitutes an essential aspect of any original composition. The literature review itself, however, does not present new primary scholarship. Some scholars believe that literature searches and subsequent syntheses of related information constitute research *per se;* however, these actions are only a part of the process. The purpose of a literature review is to glean from published works essential content for a new composition.

The review also assists the writer to describe the relationship of each work to the others under consideration. The review should identify new data and interpret the data to shed light on any gaps in previous research and resolve conflicts amongst seemingly contradictory previous studies. The areas of prior scholarship must be identified to prevent duplication of effort. This also points the way forward for further composition and demonstrates an understanding of previous related research by pointing out agreements and disagreements among the results of various studies. Additionally, the literature may suggest ways of investigating a particular problem and discuss related methodological difficulties.

Every Researchable Idea
Every researchable idea involves a number of concepts and complex relationships. In modern society, large data bases have been built around clusters of important problems. Such Internet data bases should be considered an important part of all literature searches. Research attempts to generate new knowledge. A unique synthesis of existing knowledge aids this process and is a necessary first step in any writing project, enabling an author to determine systematically what is already known. Such a search highlights questions that have not been answered. A presentation of such a review should focus on important aspects that directly relate to the objectives of the new composition.

Keep in Mind the Purpose
The purpose of a review is to assist the writer in the structuring of objectives for the new composition. This will allow the reader to be brought up to date on available literature and just how the new composition adds to or has contrasting perspectives or viewpoints than previous writings.

No doubt long lists of things to look for when reviewing the literature can be constructed. Normally, such lists are not useful. Instead, a writer should simply keep in mind the purpose for reviewing the literature. The findings will assist with refining the objectives of the new composition. What is discovered about these items is the object of the review effort.

When a problem has been identified and research questions formulated, any review of the literature for a project would include information from the body of literature by experts rather than from popular essays. Such essays may help develop a general conceptual framework for a project but cannot provide the specific theoretical basis for a new composition.

A Narrowing of Focus

All research problems fit into a larger context. A review should deal first with the general context of the larger issues and relevant theories, then, proceed logically to the specific issues. Such a narrowing of focus leads the assessor to proposed questions. Although the literature review normally follows the presentation of the problem and the purpose of the composition, it forms a foundation for the composition. Consequently, it should clearly support the objectives of the new composition.

A Review could be Endless

A literature review could be endless; consequently, one should determine the nature and limits of the literature to be reviewed. An adequate search of the literature might begin with several computerized information retrieval systems for published works and other bibliographical data bases designed to reference current published material. Most professions and virtually all disciplines have indexed important articles. The review itself should be focused on the researcher's objectives for the composition. The introduction to a new work should make it clear that the strengths and weaknesses of other research projects have been taken into account. It generally should include a current awareness search that includes citations from the previous three years in journals and five years for books in addition to any classics in the field. A good review must be more than an annotated bibliography; it ought to be a critical synthesis that prepares a foundation and structure for the new composition. Concepts and constructs used by other writers should be paraphrased using quotes only on an innovative or controlling "idea."

See Appendix Two: "Research and the Internet"

CHAPTER EIGHT

STRUCTURE AND STRATEGY

Topics Discussed in Chapter Eight

- Transparency In The Process
- Early History
- Collection Of Principles, Methods, And Strategies
- Basic Steps In Planning Social Research
- Characteristics Of Research
- Research Requires Expertise
- Elements Of A Research Study
- Hallmarks Of The Social Approach To Research
- Formulation Of The Research Problem
- Steps In The Planning And Conduct Of Social Research
- Theoretical Understanding Of The Process
- The Research Process

THEORY

INDUCTIVE LOGIC 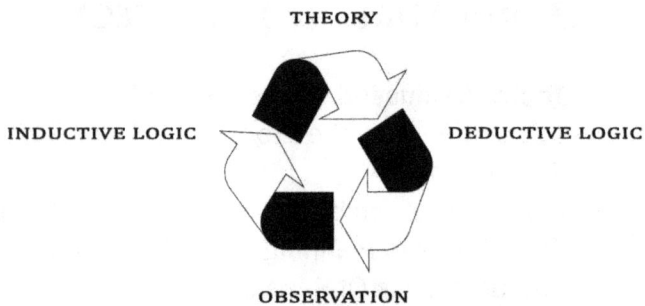 DEDUCTIVE LOGIC

OBSERVATION

The Wheel of Research

Hunch + Assumptions + Assertions + Objectives + Lit Review +
Research Hypotheses + Statistical Hypotheses + Observations +
Conclusions + Generalizations

Transparency in the Process

All reliable research is structured to provide transparency in the process of gathering and analyzing data. Those who would trust the research findings must see that standard methodology and design were employed and a feasible strategy developed based on the complexity of the research problem. Research literature offers standard and non-standard design scenarios for research. Some are too preliminary for testing hypotheses and others too expensive and time consuming to be practical. Still others require the security of a university-type laboratory.

Most social research projects, although reaching for the same results as true laboratory designs, may use less demanding approach provided the data structure and strategy relates to the complexity of the problem and the process of gathering and analyzing data is adequate and reliable. In such cases, a researcher may be permitted to design a reliable structure and strategy to fit the nature of the environment in which the research problem is identified.

Early History

In its early history, scientific research served at least a two-fold purpose; It sharpened the intellectual concerns of humans, and it assisted in developing theories to explain the universe. Human enlightenment was its goal. The earliest questions related to the universe, the grand scheme, not to such parts as the economic, social, cultural, political and anthropological aspects of human existence. Gradually, scientific prediction methods were adapted to more and more specific questions. As adaptation proliferated, disciplines emerged within the general process of scientific research. Early in its development, the lack of systematic and objective observation restricted the progress of scientific research. Thinking itself was not necessarily systematic. Overcoming this restriction required developing a way to systematically think about thinking. Thus, logic emerged.

So rooted in the earliest awareness of a need for knowledge and the continuing search by generations of thinkers, the scientific process of investigation emerge, in spite of the hurdles of vested interests and lack of support from the general populous. The concept of scientific research as understood today began to form during medieval times and developed into a full body of concepts and methodology in the modern age. This brief discussion of the development of scientific research does not include even the major contributors and precursors of modern scientific research. Research is a continuing evolutionary process set in the broader process of human inquiry

Collection of Principles, Methods, and Strategies

The following collection of principles, methods, and strategies are useful for research design and methodology. Study the material carefully. Note that most say the same thing, but differently. They are similar, but different. Although no one study would include all the items, understanding the whole process is necessary to plan and develop an adequate research projects

Basic Steps In Planning Social Research

Basic area of concern — What problem caught your interest or raised a question in your mind?

Rationale and theoretical base — Can this concern be fittcd into a conceptual framework that gives a structured to the research? In other words, can you begin from a position of logical concepts, relationships, and expectations based on current thinking in this area and build a conceptual framework providing definition, orientation, and direction to your constructive thinking?

Purpose/problem statement — Define the problem and state the objectives. What aspect of the larger problem will you research?

Ultimate questions — When the research is finished, what are the questions to which reasonable answers can be expected?

Statement of objectives and/or hypotheses — State the specific objectives of the research and the hypotheses you will test. Be concrete and clear, making sure that each objective or hypothesis is stated in terms of observable behavior allowing objective evaluation of the results.

Design and procedure — State who the subjects will be, how they will be selected, the conditions under which the data will be collected, treatment variables to be manipulated, what measuring instruments or data-gathering techniques will be used, and how the data will be analyzed and interpreted.

Assumptions about the problem — What assumptions have you made about the nature of relationships that may have precipitated the problem?, What about methods and measurements, or about the relationship of this study to other persons and situations?

Operational terms — List and define the principal used in hypotheses, particularly where terms have different meanings to different people. Emphasis should be placed on operational or behavioral definitions.

Scope and limitations – What are the limitations surrounding the study and within which conclusions must be confined? What limitations exist in the methods or approach–sampling restrictions, uncontrolled variables, faulty procedures, etc.? How has the scope of the study been arbitrarily narrowed? Was focus only on selected aspects of the problem, certain areas of interest, a limited range of subjects, and level of sophistication involved?

Instrumentation – Will you construct an instrument or use one from another source? What other compromises to internal and external validity may exist?

Characteristics of Research
Research is directed toward the solution of a problem, emphasizes the development of generalizations, and is based on observable experience. Research demands accurate observations and descriptions and involves gathering new data from primary sources or using existing data for a new purpose.

Research requires Expertise
The researcher must know what is already known about the problem and how others have investigated it previously. Research strives to be objective and logical, applying every possible test to validate the procedures employed, the data collected, and conclusions reached. Research is characterized by patience and unhurried activity and is carefully recorded and reported. Research often requires courage.

Elements of a Research Study
The elements of scientific methodology provide the basic ingredients of a research study. They may be listed as: 1) problem identification, 2) a constructed hypotheses, 3) design of the study. 4) data collection, 5) data analyzed and

interpretation, 6) and the results reported. Another way to list the elements would be: problem identification and hypotheses formulation, research structuring, methodology selection, data collection, hypotheses testing, and research reporting.

Hallmarks of the Social Approach to Research

These characteristics exhibit the basic nature of the social approach. One must always stress the objective, systematic, and control nature of the scientific process. This enables others to have confidence in research outcomes. The fundamental of the scientific approach is a controlled rational process. While only an ideal science would exhibit each of the following six characteristics, they nevertheless are the hallmarks of the scientific approach to research.

The procedures are public. This means that scientific research contains a complete description of what was done, to enable other researchers in the field to follow each step of the investigation as if they were actually present.

The definitions are precise. The procedures used, the variables measured, and how they were measured must be clearly stated.

The data collecting is objective. Bias in collecting data, as well as in the interpretation of results has no place in science. Objectivity throughout is the key feature of the scientific approach.

The findings must be replicable. This enables other researchers in the field to test the findings, or results of the study by attempting to reproduce them.

The approach is systematic and cumulative. This relates to one of the underlying objectives of science --to develop a unified body of knowledge.

Research objectives include understanding, explanation, and prediction. Every scientist wants to know the how and why. If one determines the **how** and **why** and is able to statistically support an antecedent cause, then prediction is possible. Prediction is the ultimate objective of science.

Formulation of the Research Problem
A question well stated is a question half-answered. If one asks an intelligent question, this person has the capacity to find an answer.

A group capable of identifying a problem can find a workable answer. One's thinking may be part of the problem, or it could be part of the answer. It is easy to make mistakes in problem formulation.

Some common mistakes are: Collecting data without a well-defined plan, hoping to make some sense out of it afterward. Taking a "batch of data" that already exists and attempting to fit meaningful research questions to it. Defining objectives in such general or ambiguous terms that your interpretations and conclusions will be arbitrary and invalid.

Undertaking a research project without reviewing existing literature on the subject or doing ad hoc research, unique to a given population, permitting no generalizations beyond the subjects and making no contribution to the general body of research.

Other failures in problem formulation process are: Failure to base research on a sound theoretical or conceptual framework, which would tie together the divergent masses of research into a systematic and comparative scheme, providing feedback and evaluation for theory. Failure to make explicit and clear the underlying assumptions within one's research so that it can be evaluated in terms of these foundations. Failure to recognize the limitations in the approach, implied or explicit, that place restrictions on the conclusions and how they apply to other situations. Failure to anticipate alternative or rival hypotheses that could also account for a given set of findings and which challenge the interpretations and conclusions reached by the research.

Steps in the Planning and Conduct of Social Research
Identify the problem area and survey the literature relating to it. Define the actual problem for investigation in clear, specific terms. And formulate testable hypotheses that define the concepts and variables. Then state clearly the underlying assumptions, which govern the interpretation of results and develop the research design to block for all threats to internal and external validity.

Now you are ready for the selection of subjects by determining the population and the sample size. Next choose or construct an instrument and make certain it has been properly validated. Then specify the data collection procedures and establish the criteria to evaluate the outcomes or test the hypotheses, that is, what statistical procedures will be used. How will the data be analyzed and conclusions be made. Finally, what report form will be appropriate for the research.

A Respected American Sociologist
I discovered the work of Earl R. Babbie in 1977. He is a respected and influential writer in the field of social research. As a sociologist and professor in the Behavioral Sciences, his work has provided assistance to many teachers and students. Following are two diagrams that

demonstrate Babbie's theoretical understanding of the research process. These diagrams and the twelve editions of *The Practice of Social Research* were widely used in my classroom for three decades as a Professor of Education and Social Change. It is my firm belief that no one better understands social research than Dr. Babbie.

Babbie's Theoretical Understanding of the Process

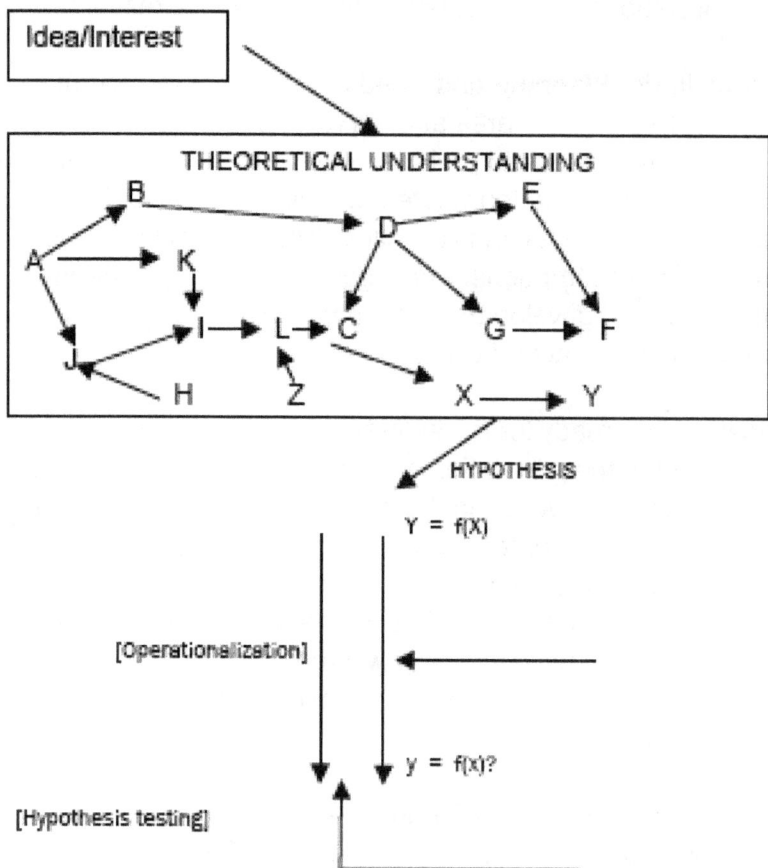

The Traditional Image of Science

Babbie's Elaboration of the Research Process

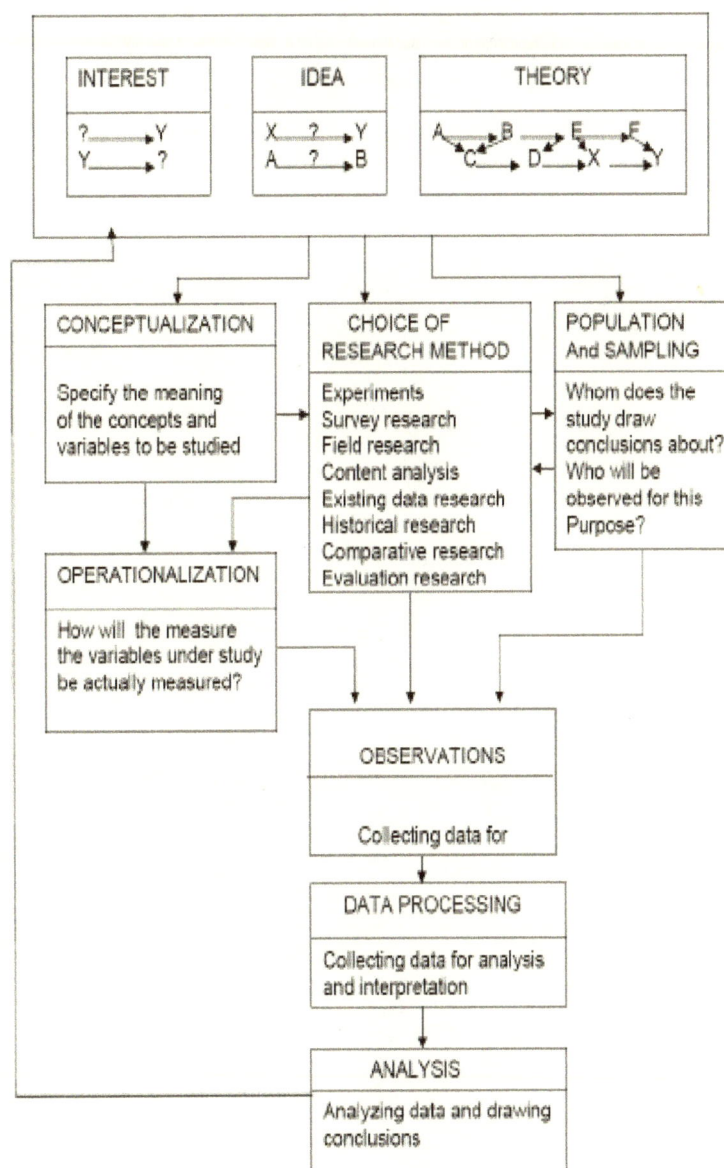

INTEREST

?————▶Y
Y————▶?

IDEA

X——?——▶Y
A——?——▶B

THEORY

A———▶B———▶E———▶F
C———▶D———▶X———▶Y

CONCEPTUALIZATION

Specify the meaning
of the concepts and
variables to be studied

CHOICE OF
RESEARCH METHOD

Experiments
Survey research
Field research
Content analysis
Existing data research
Historical research
Comparative research
Evaluation research

POPULATION
And SAMPLING

Whom does the
study draw
conclusions about?
Who will be
observed for this
Purpose?

OPERATIONALIZATION

How will the measure
the variables under study
be actually measured?

OBSERVATIONS

Collecting data for

DATA PROCESSING

Collecting data for analysis
and interpretation

ANALYSIS

Analyzing data and drawing
conclusions

CHAPTER NINE

METHODOLOGY AND DESIGN

Topics Discussed in Chapter Nine

- Controlled Laboratory Required
- Standard Methodology
- Purpose/Problem Statement
- Qualitative And Quantitative Research
- Exhaustive Logic And Literature Review
- Support For Conclusions
- A Few Statistics
- Historical And Theoretical
- Value And Accuracy
- Academic Background
- True Laboratory Designs
- A Virtual Laboratory Design
- A Matter Of Scaffolding
- Alternative Explanations
- The Same Results
- The Practice Of Intentionality
- Operational Definitions
- Surrogated Assessment

THEORY

INDUCTIVE LOGIC

DEDUCTIVE LOGIC

OBSERVATION

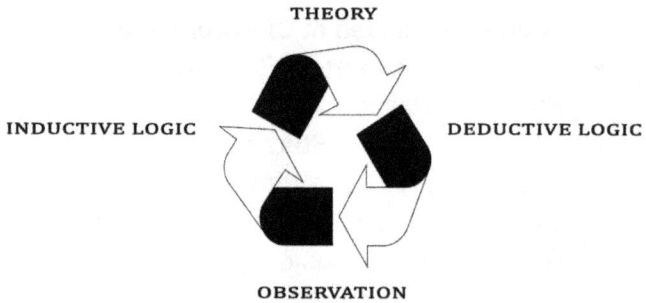

The Wheel of Research

Hunch + Assumptions + Assertions + Objectives + Lit Review +
Research Hypotheses + Statistical Hypotheses + Observations +
Conclusions + Generalizations

Controlled Laboratory Required

The occurrence or cause for a particular outcome, a secure, controlled laboratory is required. An antecedent cause is determined by answering the question as to why an unusual fact or occurrence happened. A true experimental research design is normally required to determine an antecedent cause. The ability to generalize the findings however is limited to the subjects of the laboratory experiment. Since a goal of social research is to generalize findings to a larger population from a sample, all reliable research is structured to provide transparency in the process of gathering and analyzing data.

Standard Methodology

It is necessary that research be placed in the framework of standard methodology to assure trust in the findings. The choice of methods is based on the complexity of the research problem. Literature provides an assortment of research designs. Choose the design that fits the research and provides the best chance of testing the proposed hypotheses. Still others require the security of a university-type laboratory. Most social research projects, although reaching for the same results as true laboratory designs, may use less demanding approach provided the data structure and strategy relates to the complexity of the problem and the process of gathering and analyzing data is adequate and reliable. In such cases, a researcher may be permitted to design a reliable structure and strategy to fit the nature of the environment in which the research problem is identified.

Purpose/Problem Statement

When the research problem is identified, a purpose or problem statement should be developed to indicate precisely which aspects of the problem will be examined. Most social problems are too complicated to consider as a whole; therefore, the background of the problem may be an hour-glass approach that identifies the general problem and the various aspects of the problem that are visible in society. Notwithstanding, the problem identification must culminate in a description of the portion of the problem the research will observe. The observation strategy must be articulated to include assertions of either theoretical contributions or practical solutions related to the specific problem. Once the structure and strategy for this process is clearly delineated, the researcher may determine the contribution to scientific understanding the research will make. This prepares the researcher to support a significance statement related to the research outcomes.

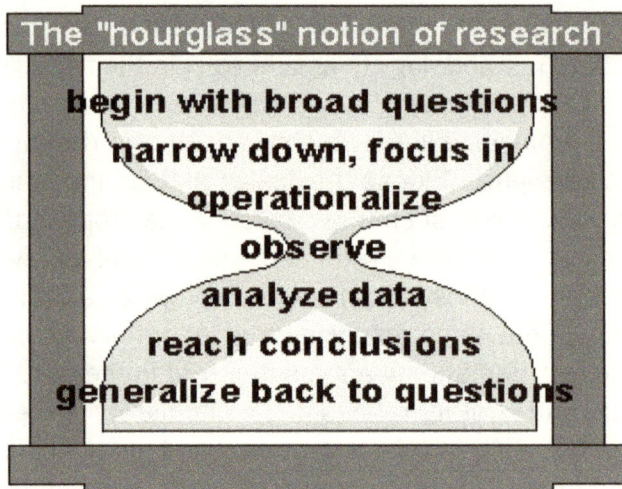

The "hourglass" notion of research

begin with broad questions
narrow down, focus in
operationalize
observe
analyze data
reach conclusions
generalize back to questions

Credit for the hourglass drawing goes to
William W. Starkey

A purpose/problem statement must be developed early in the process to indicate precisely which aspects of the problem will be examined. Most social problems are too complicated to consider as a whole; therefore, the background of the problem may be an hour-glass approach that identifies the general problem and the various aspects of the problem that are visible in society. Notwithstanding, the problem identification must culminate in a description of the portion of the problem the research will observe. The observation strategy must be articulated to include assertions of either theoretical contributions or practical solutions related to the specific problem. Once the structure and strategy for this process is clearly delineated, the researcher may determine the contribution to scientific understanding the research will make. This prepares the researcher to support a significance statement related to the research outcomes.

Qualitative and Quantitative Research
Qualitative and quantitative research are both necessary to accomplish social scientific research. Qualitative research is necessary to uncover the problem and analyze previous research on the subject and quantitative research is necessary to test the assumptions and hypotheses formulated on the subject. The two methods of research support one another and are not in opposition, nor is one considered more scholarly than the other. The precise nature of the research problem or question determines whether research will be predominantly qualitative or quantitative in terms of the research methodology. Each method requires a distinct approach to the solution of the problem.

The nature of the research design will determine whether the emphasis or weight of the research project falls on one side or the other of the quantitative -- qualitative dichotomy. Quantitative research is concerned primarily with measuring or numbering a research hypothesis to determine its quantitative significance. Qualitative research is meant to signify that the research problem or question is being studied with reference to its historicity, essential ideas, or unique characteristics. Students should understand the nature and importance of each method and how both fit into the research process. A decision as to which method to stress should not be made until the candidate develops a full research proposal.

Exhaustive Logic and Literature Review
Qualitative research requires exhaustive logic and literature documentation to support each aspect of the research. In quantitative research the Literature Reviews is comprehensive, but not exhaustive. The literature places the problem in the context of current research and normally becomes Section Two of the research proposal with little if any references necessary elsewhere. It is a comprehensive review of published material related to the hypotheses proposed for the research and a current awareness search in journals for the past three years. A rigid design and statistical process supports the conclusions.

In qualitative research, an exhaustive literature review must deal with the whole subject area as background support for the problem in the present study. For example, a qualitative study on a religious denomination dealing with a particular social or doctrinal issue would require an exhaustive review of the history of the denomination as it related to the problem at hand. Those who do qualitative dissertations normally have writing and publication experience, bachelors and masters degrees in the subject field, and/or extensive doctoral courses in the content area. One cannot jump into a qualitative study without an exhaustive effort to place the problem in the historical context of the environment under study. Finding primary sources is often the greatest difficulty.

Support for Conclusions

One major concern in the qualitative area deals with the conclusions. How does one support the conclusions and exclude rival explanations? This requires exhaustive work. Seminaries, some departments in universities, and some doctorates of lesser quality than the PhD permit qualitative research, but universities that offer the PhD-type degree prefer scientific research that includes statistical testing of hypotheses and support for the findings. The better universities that accept qualitative dissertations normally require logic chains and Boolean logic in support of conclusions. The Boolean approach is a mathematical system originally devised for the analysis of symbolic logic, in which all variables have the value of either zero or one. It is this process that is used to support the digital computer system.

A Few Statistics

Research supported by quantitative means is normally much shorter than qualitative research. Supporting conclusions by logic and Boolean algebra requires critical thinking, sound reasoning, and good journalistic skills. A few statistics can save many written pages and provide almost instant support for the conclusions made in a quantitative study. For example, qualitative research needs a preponderance of the evidence (perhaps 95%) to support reasoned conclusions. These may be supported by Boolean algebra (logic) which is based on reason and logic and is a long process. While quantitative research can reject a null hypotheses at the statistical .05 level of confidence and consequently support the original research hypotheses. By going in the back door and finding a small relationship or difference, the null hypothesis may be rejected; this automatically supports or confirms the original assumption or hypothesis. This could mean the difference in 25 pages of qualitative writing versus 5 pages of support for one hypotheses or conclusion is supported by statistical manipulation.

Historical and Theoretical

Qualitative research is historical and theoretical in nature and indicates a study of primary and secondary sources that recount the past history and narrates the present context of the research topic. Theoretical research is a study of sources in order to discover, explain, formulate, compare, or analyze theories or ideas associated with the research topic. In historical and theoretical research, it is crucial to have authentic and accurate research data. This data comes primarily through secondary and tertiary sources.

Value and Accuracy

The sources are used to determine the value and accuracy of any research assumption or conclusion. The veracity of these sources comes through external and internal criticism. External criticism determines the authenticity of the sources. Internal criticism evaluates their accuracy or worth. Once the credibility of the primary and secondary sources is established, the researcher critiques these sources in relation to the particular problem or question under study. Both historical and theoretical research is to be as exhaustive as possible.

Academic Background

One of the reasons students are directed to social research rather than historical studies, is the content and academic background is not normally there to take the qualitative approach. It is not that qualitative research is substandard; it is mainly that statistically supported research is usually easier for the inexperienced researcher. Doctoral candidates are normally writing a dissertation as a first effort at a major project. One wants conclusions to be accepted and statistical support for conclusions are normally more readily accepted by both academia and the organizations and institutions related to the research problem. Often, without exhaustive documentation in support of conclusions, the reasoning or logic can be easily questioned.

Academics are not opposed to this kind of study, but the methodology is difficult to teach because it is learned through trial and error and experience. Also, the length of time is exhausting and the final data less appreciated. When students make mistakes in reasoning and logic in a short research papers, one can imagine what could happen in a long qualitative research project.   In my judgment, statistics is much easier to learn than Boolean algebra and/ or logic. The thought is enough to make one want to learn statistics.

True Laboratory Designs
Research requires both control and comparisons.   To discover antecedent cause a true research design is required. When a researcher is not searching for a preceding cause, quasi-experimental research designs are formulated. To accomplish this task the social researcher must step outside the comfort of a university research laboratory and create a virtual laboratory that imposes both sufficient control and adequate comparisons.

A Virtual Laboratory Design
A university research laboratory is designed for scientific experimentation in a secure and protected space. The security measures imposed on a research laboratory determine the reliability of the experiments. The work space is protected from the inclusion of all unnecessary or insidious (sinister, dangerous, subtle) elements that would contaminate the experiments or invalidate the process. To create a virtual research laboratory, the researcher must

have a clear understand of all threats to the validity or reliability of the intended research and construct a practical virtual laboratory that is equivalent as far as effect is concerned. The threats to validity are addressed through the internal virtues or powers of the virtual design. The virtual laboratory is a research lab without walls.

A Matter of Scaffolding

The construction of a virtual laboratory is a matter of scaffolding a temporary platform using standard research planks so the researcher may clearly see what needs to be accomplished. The virtual lab enables the researcher to adequately utilize the resources and tools available to complete the research task in an acceptable manner. The problem areas must be clearly seen, described, and an applicable solution process applied. The scaffolding is a visible platform upon which the researcher may gain perspective and work to describe the problem, gather and analyze data, and report precisely the findings. The scaffolding clearly is a structure to support the research task and a platform to view the problem clearly and apply appropriate and acceptable methods to the task at hand.

Alternative Explanations

Research designs require that most rival hypotheses (alternative explanations of the covariance) be controlled so that the causal relationship hypothesized can, in fact, be tested. In the absence of such controls, the covariance that may be attributed to rival hypothesis may swamp (obscure) the covariance attributable to the research hypothesis. Such controls may be ingeniously designed and unique to a particular research project. However, several designs have broad application have been developed. The concept of experimental control is illustrated with such designs. In designs of this nature, statistical blocking (including extraneous variables in the mathematical model) and random error (a residual variable that captures the variation attributable to all non-specified experimental and extraneous variables) are used to enhance the control procedures.

In virtual laboratory designs, the diagram or box stands for the research sample, "E" represents the experimental manipulation, "A" is an attitude assessment, "K" is a test of knowledge, and "B: is a behavior measurement. We use these three specific designators in place of the more general "O" that is commonly used to designate observations or measurements of any sort. The arrow ($\rightarrow$)

Virtual Laboratory Design

COMPARISONS AND CONTROL

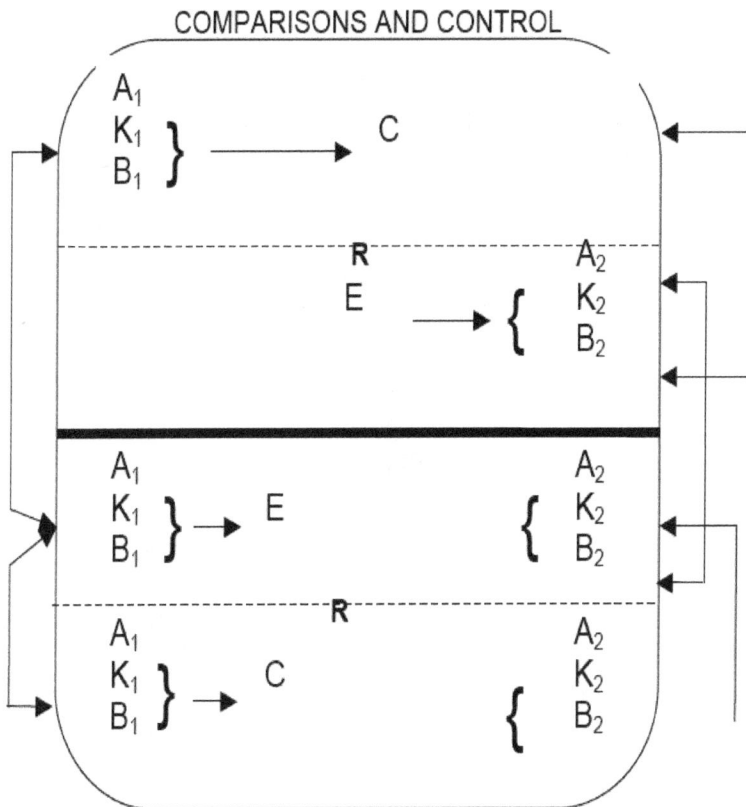

means "followed by." The arrow alone indicates that an assessment or measurement is made immediately following the experimental manipulation. Delayed assessment or measurement is indicated by and arrow super-scribed with a "(T)," i.e., $-^{(T)}\rightarrow$. Superscript $^{(T)}$ indicates a point in time. The broken line is used to divide the box, representing the sample, into portions corresponding to the number of groups into which the sample is divided. The "R" indicates that all subjects are randomly assigned to the groups. A subscript "1" identifies a pretest and the subscript "2' denotes post-test. The Virtual Laboratory Design is an example of how a true experimental design known as the **Solomon Four-Group Design** may be adapted for a virtual laboratory identified as a **Split-intact Pretest/ Post-test Design**.

The Same Results

Most social research projects, although reaching for the same results as true laboratory designs, may use less demanding approach provided the data structure and strategy relates to the complexity of the problem and the process of gathering and analyzing data is adequate and reliable. In such cases, a researcher may be permitted to design a reliable structure and strategy to fit the nature of the environment in which the research problem is identified.

The Practice of Intentionality

Consider your left hand and your right hand as a philosophical construct and mechanism to divide the process of research. The left hand is considered "we – group action" and the right hand understood as "me – individual action." The ring finger is on the left hand; therefore, the wedding band is normally worn on the left hand because it signifies a "we" relationship with another. Whether it is politics – left and right, social and spiritual change, contextualization and ortho-praxis, or many other dichotomies, the left and right construct may be used to better understand the process. The difference can be explained in terms of "we – the group action" or "me – individual action.

Operational Definitions

A term such as "religious vocation" must be defined in words that are objective and that would result in all persons who use the definition making identical decisions about any particular case. This process—**operationalizing**—is not an easy process. Operationalization became prominent in the social sciences in the 1920s when its disciplines were striving to compete with the "hard" (or "exact") sciences by adopting rigorous quantitative methods. Operationalization holds that the validity of a given research finding or theoretical construct is dependent on the validity of the operations used in the process. In other words, one cannot produce valid results without a disciplined process. What is the unit of measure for anger, love, or compassion? The solution was not to study the abstract concept directly, but to study behaviors by which the concept was believed to be manifest. This is a surrogating process. It is only by observing and measuring specified behaviors—called operational definitions—that abstract concepts can be used in social research.

A Faith-based Perspective

Faith-based groups have always maintained that truth exists and is knowable, but recognizes that some varieties of human experience cannot be measured directly. Hence, answers to certain questions—some specific truths—will not be found by the use of empirical processes. Ancient faith-based writings offer revelation as a means of gaining otherwise inaccessible truth, and holds that some truths are knowable only through faith. Validate or reject this assertion.

Does scripture use operational definitions, or at least authorize their use? Examine I Timothy 6:3-5. Might these be considered operational definitions of a teacher of false doctrines? See Proverbs 31:10-31 for operational definitions of "the wife of noble character." A search of scripture would discover other operational definitions.

Surrogated Assessment
An indirect approach, called surrogating, may be necessary when behavior cannot be measured directly. Since no two researchers agree on the mental process of others; therefore, un-observables behavior may be assessed by indirectly determining the predisposition to act in a given situation. With an index of questions, such case may be assessed as above or below the mean; therefore, an indirect measurement has occurred. Sensations and perceptions may not be measureable by standard methods, but one's predisposition to act may be assessed. The concept of sensation or perception may be evaluated using a surrogated measurement.

It is crucial that designers of social research understand the difference between behavior and attitudes or knowledge. An attitude is a predisposition to act and requires an assessment rather than a direct measurement. The level of knowledge may be established, normally by a pre-test and a post-test or some other evaluation criteria. Since behavior is a goal directed activity, "behavior" may be measured directly. The key to designing valid research is to know the difference in the kind of measurement required to obtain valid conclusions.

LAYOUT AND COMMUNICATION

Topics Discussed in Chapter Ten

- Problem Identification
- Hypotheses Formulation
- Operation Definitions
- Scope And Limitations
- Significance Statement
- Research Methods
- Instrumentation
- Statistical Procedures And Research Design
- Measurement And The Assignment Of Numbers (Values)
- Assignment Of Numbers
- Levels Of Measurement
- Checklist For Research Development

THEORY

INDUCTIVE LOGIC 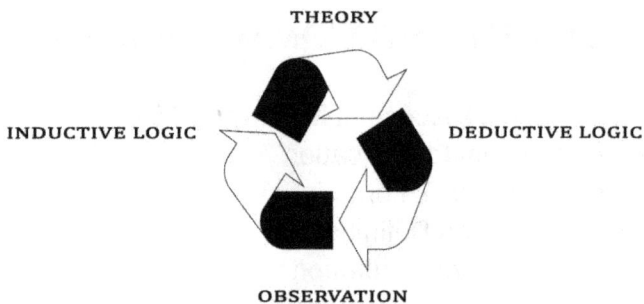 DEDUCTIVE LOGIC

OBSERVATION

The Wheel of Research
Hunch + Assumptions + Assertions + Objectives + Lit Review +
Research Hypotheses + Statistical Hypotheses + Observations +
Conclusions + Generalizations

P roblem Identification

Although research does attempt to push back the horizons of knowledge generally, specific projects are often couched in the context of specified problems. An adequate proposal includes identification of a problem. The background literature concerning a problem is used to present trends, unresolved issues, social concerns, and the like. Literature study often makes it possible to limit a research project to some aspect of a particular problem. By focusing on specific aspects and viewing such aspects in the context of related considerations, researchers delimit the purpose of a particular research project. A purpose statement indicates precisely which aspects of a problem will be examined. Variables and major research questions to be considered should be identified and stated clearly. These questions are based on assumptions about the research problem and a literature review based on the objectives of the research. This process creates the theoretic framework for developing both objectives and hypotheses.. Hypotheses are formulated so that test results will answer the major research questions.

Hypotheses Formulation

Hypothesis is defined as an assertion that can be confirmed or disconfirmed by empirical evidence using understood methodology and methods. A common misconception is that research questions are really quite general and perhaps a little vague. Although research hypotheses are quite general, they are certainly not vague. Actually, both specificity and clarity must be accomplished in the research question step which connects the research project to the extant body of scientific knowledge. This connection must be clear and specific. Research hypotheses restate clearly and specifically the stated research questions into assertions that if accepted answer clearly and specifically those questions. Statistical hypotheses finally restate research hypotheses into the terms of specific statistical logic systems so that the hypotheses may be tested quantitatively.

Hypotheses serve a number of roles in research. They suggest the general form the research design must take, including the type of data needed, sorts of instruments needed, and statistical procedures likely to be used. Their direct and most vital function in the research plan is to provide a formal connection between research questions arising from the study of existing knowledge and conclusions about those questions. They provide the logic by which new empirical evidence is introduced to the already understood mass of scientific information. The progression is clear: the research question is based on the scientific literature. The research hypotheses is based on the research assumptions. The conclusions are based on the statistical hypothesis.

Operation Definitions
Once hypotheses are clearly stated, definitions of important terms used in the hypotheses should be provided to clarify their operational or technical use. Many words used in research have several common meanings. If such words are used as terms of a hypothesis, the operational meaning for the present research is used. Occasionally, no standard term exists to describe to make as fine a distinction as a researcher intends. In such cases, a precise operational definition should be made. However, a caveat should be emphasized with reference to describing a more precise operational definition than their research is able to make. As discussed throughout the text, scientific instruments, measurement theory and methods, and statistical inference logic limit present ability to observe phenomena. An investigator cannot, simply by providing a name, make

something a particular thing. Precise definitions help
the researcher to interpret the research questions and
hypotheses. Operational definitions should include how the
term is being measured or manipulated in the research.
Technical terms needing explanation, but not a part of the
hypothesis, should be placed in a glossary.

Scope and Limitations

Most modern problems are multifaceted and complicated.
It is unrealistic to expect that any one research project will
totally solve such problems. Consequently, all research
projects have limitations. The scope of a research project
should be clearly focused on a problem and a population
with the known limitations identified. Limitations affect
both the interpretation and the ability to generalize the
findings. Limitations are discovered throughout most
of the process of developing a proposal. By the time a
proposal is written for presentation, a researcher clearly
understands the limitations of the proposed project.
Occasionally, a limitation may be overlooked in the planning
stage. The impact of such a failure can range from minor
to catastrophic, depending on how the overlooked limitation
relates to various elements of the research project.
Research is a process. A clear description of the conceptual
framework that guides a study together along a logical path.
By distinguishing a particular path from inappropriate ones,
a researcher defines the scope of a project.

Time resources, available data, and the like may limit
a particular project. Additionally, conceptual and
methodological shortcomings that are known, but cannot
be overcome, should be disclosed. Such shortcomings
certainly should be minimized. Every shortcoming is a
potential basis for rejecting the results of the study. While
overtly acknowledging such shortcomings is some degree
better than not doing so, their disclosure does not change

their character. They are shortcomings, weaknesses in a study.

Significance Statement
All research projects should be positioned in the context of other studies that may concern the same or similar subjects. What similarities and distinctions can be drawn between a particular study and related ones? In this framework, a significance statement discusses the importance of a particular research project based on conclusions from the findings.

This section of a proposal describes the utility of the focus chosen by the investigator. It should contain adequate justification for the research beyond a desire to publish the findings, establishing that the approach is appropriate and that the study will benefit certain individuals, groups, or organizations. A significance statement may include assertions of either theoretical contributions or practical solutions of specific problems. It should clearly state what contribution to scientific understanding the study will make.

Research Methods
The methodology section of a proposal is a blueprint of the procedures to be followed to answer the research questions and to interpret the results. Although most of the terminology associated with research design originated with experimental research, all research must have a design that clearly integrates the process and structure of the study.

Populations that concern the project should be clearly identified. Methods by which the researcher intends to select participants for the study and sampling plans should be described and detailed sampling procedures specified. The description should include the size of any samples, and how subjects are assigned to samples, whether the subjects will be volunteers or be required to participate.

Characteristics of populations constrain samples. The natures of both samples and population affect the ability to generalize the results of any study. No results can be generalized beyond the population from which the sample is considered to be representative. An adequate sample must be representative and independence. Results can be generalized to the population only when the sample is representative. Independence of individuals or groups within the sample is required by most of the statistical procedures used to analyze research data. Both being representative and independence are provided by the major sampling procedures: random sampling, stratified random sampling, cluster sampling, and, in some cases, systematic sampling. When systematic sampling is used, the researcher must be careful to see that the predetermined list contains no concealed periodic bias that would produce an unrepresentative sample.

Sameness or (representative) can also be jeopardized by individuals who refuse to participate in the study and by volunteers. Consequently, non-responses can be a serious problem in studies using mailed questionnaires. Non-response problems can be minimized by providing replacements through special procedures and by following up initial questionnaires with reminders to non-responsive individuals. Using volunteers may limit the ability to generalize the findings. Avoiding using volunteers and documenting procedures to insure full participation of individuals sampled strengthens this process.

The individual sample should be large enough to minimize sampling errors and to provide adequate power for the statistical procedures used in data analysis. The power of a statistical procedure is the possibility that the procedure will adequately detect significant effect. Power is defined as the probability of rejecting the null hypothesis, given that it is false. Power depends upon the level of significance, the size of the effect, and the size of the sample. The level of significance is defined as the probability of falsely rejecting a value of .05 or .01. The size of the effect represents the difference that actually exists between parameters of interest as a ratio. If the level of significance and the effect size are fixed, power varies directly with the sample size. The larger the sample, the higher the power. (A complete treatment of sampling procedures and sampling size projections are found in most introductory statistics textbooks.)

Instrumentation
The methodology section should include a description of measuring instruments or procedures to be used to provide data for the study. Each should be described in detail. Will a questionnaire or a structured interview be used? How is the instrument organized? The instrument or procedure used in a research must be both valid (measure what it purports to measure) and reliable (produce consistent scores). This section should include evidence for the validity and reliability of any instruments or procedures used, including types of validity and reliability and sources for the evidence. If possible a copy of the instrument should be appended to the proposal.

Effort should be made to use existing instruments. There are several advantages to this. Considerable time can be saved. Instruments that have been used in previous research normally can provide sufficient reliability and validity data to permit evaluation of appropriateness and to make the use of the instrument easier to defend. Also, results can be compared to those of others who have used this same instrument. Instruments cited in journal articles, research papers, and dissertations are usually widely available. If not, they can normally be acquired from the author of the research.

When an appropriate instrument cannot be located, the researcher must construct a new instrument. This is time consuming and requires field testing and careful analysis of results before the instrument can be used in research. Despite the work, it is better to develop a new instrument than to attempt to use an extension instrument that is not entirely appropriate. (Consult textbooks on test construction for guidance and a more detailed explanation of validation procedures.)

Statistical Procedures and Research Design
Statistical procedures associated with the research design being used should be described, together with the procedures proposed for data analysis. Evaluators of a proposal should be given details of statistical designs and procedures adequately to evaluate their appropriateness and the researcher's ability to execute the study. Such details may also benefit other researchers who wish to build on a study. The carefully prepared set of procedures is an invaluable guide to the actual execution of the research.

A written narrative accompanied by a schematic diagram may be used to present a research design and to show the relationship of a particular statistical procedures to the design. Presentation of statistical procedures should include the name or description of each procedure and of the dependent and independent variables, the level of significance, and a description of how particular procedures connect to particular hypotheses and research questions. The appropriateness of statistical procedures depends on logical connecting procedures to hypotheses.

If possible, it is best to use procedures that are widely known and commonly employed by other researchers. Academic courses and adequate literature review in the field provide a basic understanding for selected statistical procedures appropriate for selected research designs. We have provided references to lead students into the statistical literature at the end of chapters that discuss such procedures.

Measurement and the Assignment of Numbers (values) Measurement relates to any devise used to find and express size, extent, or capacity. Measurement is an important set of procedures for insuring the comparability of various scientific observations. Measurement theory can be broadly categorized into classical theory and modern theory. Classical theory requires that numbers be assigned to observations in a manner that is isomorphic to the numeric system termed *arithmetic*. Modern theory allows isomorphism with other numeric systems.

Modern measurement theory allows the construction of various levels of measurement scales. Such scales have proven useful for research in the social sciences. However, methods of assigning numbers to observations are not always as rigorous in terms of system behaviors as they should be. as a result, currently all measurements used in scientific research are not necessarily comparable.

Classical Theory is based on the science of numbers (arithmetic) –the fundamental operations in arithmetic are addition, subtraction, multiplication, and division. Arithmetic deals with counting real numbers.

Modern Theory relates with the group of science including arithmetic, geometry, algebra, calculus, etc., dealing with quantities, magnitudes, and their relationships, attributes, etc. by the use of numbers and symbols. Modern Theory requires more rigor and exactness, precision, and accuracy. Modern Theory deals with the science of quantities and magnitudes, the relations between them, and the methods by which unknown quantities can e found from those known or supposed, expressed, or calculated in numbers or other symbols.

Assignment of Numbers
Numbers are first assigned according to the empirical relationships discovered by observation and then rules are constructed delimiting the operations that can be performed on the assigned numbers. By applying this methodology social scientists are able to map significant properties of groups, organizations, and societies. Most social scientist accept the extension of measurement to include structures of methods of assigning numbers to observations that are isomorphic(same) to numerical systems that are weaker than the arithmetic system.

Different statistical procedures are used with different data: nominal, ordinal, interval, or ratio. The selection is down only from nominal to ratio, nominal being the weakest form of data and ratio being the strongest data. Statistical assessment is used to describe the actions that assign numbers to certain non-transformable systems. Consequently, physical behaviors are measured while judgments, attitudes, and intuitions are assessed.

Distinguishing between measurements and assessments emphasizes the connection between a concrete system (existing in physical space-time) and a conceptual system (statistical space). In a general sense, it is useful to define measurement as the formal comparison of two regions of space-time, (1) the scale and (2) the object of measurement. The "formal" denotes accepted methodology, both the scale and the object of measurement must be part of the universe of concrete systems. The scale is a previously agreed-upon quantifiable conceptual system, e.g., a ruler, pounds, liters, Fahrenheit degrees, miles per hour, categories, ranks, and the like. That is to say, it is a system of information and, as such, it must be borne on information markers (special bundles of matter/energy) in the universe of concrete systems. An observer first determines that the relationships of a scale represent the relationships of certain elements of concrete systems and then uses the scale, assigning the numbers in this conceptual system to elements of the concrete system, to create a unique conceptual system that represents the concrete elements. This system of information can then be used to study the elements of the concrete system.

Levels of Measurement

Levels of measurement are distinguished one from another by certain formal properties. Mathematicians and statisticians have formulated these properties precisely with symbolic axioms that identify the operations of scaling and the relationships among the scaled objects. While these axioms are fairly exact definitions of the nature of the scales, assessing the level of measurement achieved by a set of numerical data. Because they are widely accepted, the nominal, ordinal, interval, and ratio levels of measurement are discussed. Each higher level subsumes the properties of those below it. For guidance in choosing the proper statistical procedure consult the Suggested Reading and Reference section.

A **nominal scale** (also called a classificatory scale) is created when numbers or other symbols are used to identify (name) the groups to which objects are assigned. Such assignment constitutes the weakest form of measurement and is thus viewed as the lowest level.

An **ordinal scale** (sometimes termed "ranking scale") is
created by using numbers or other symbols to identify the
groups to which objects are assigned in a manner that
signifies some kind of relationship among the groups.
Among relationships are: more, less, higher, and lower.
On this scale, the equal objects within groups are not just
different from those of other groups but also hold a specific
relationship to each of the other groups. Assignment of
numbers in this manner adds the quality of relationship to
those of the nominal scale, and thus, it achieves a more
precise or sophisticated level of measurement.

An **interval scale** is created when numbers or other symbols
are used to identify groups to which objects are assigned
so that the relationship among the groups is specified by
a common and constant unit of measurement. That is,
the distances between any two groups are known. This
is an extension to all numbers on the scale itself of the
requirement of lower levels that objects within groups be
equivalent units in terms of the scaled property. Thus, the
equivalence rule applies to both within and between group
measurements. This is the first truly quantitative scale.

A **ratio scale** is created when numbers or other symbols
are used to identify groups to which objects are assigned
in such a way that the relationships among the groups are
specified by a common and constant unit of measurement
and the scale of these units has a true origin. The only
difference between the interval and the ratio scales is that
zero has true meaning (the absence of the property being
measured) on the ratio scale and this is not so on the
interval scale.

One must know the level of measurement before a proper
statistical procedure can be selected. There must be a
goodness of fit or the conclusions will not be valid. Certain

statistical procedures are used primarily with specific levels of measurement; such as, nominal, ordinal, interval or ratio data.

See Appendix Three: Non-parametric Statistical Procedures

Checklist for Research Development

Not all of the items apply to any one research proposal. Items needed are determined by the topic and research methodology. Valid designs have several features:

The Problem

_____Problem to be studied is identified and stated

_____Background (e.g., trends, unresolved issues, social concerns) of the problem is presented

The Study

_____Purpose of the study is stated, emphasizing outcomes or products

_____Nature of study is described

_____Conceptual or substantive assumptions are stated

_____Rationale and theoretical framework for problem solution stated

_____Hypotheses (objectives) to be researched are stated

_____Importance of the study (so what?) is set forth (may overlap with statement of problem)

_____Operational terms are defined

_____Scope and limits of the study are specified

Literature Review

_____Historical background of problem is developed

_____Reader is acquainted with existing related studies

_____What has been found?

_____Who did the work? When? Where?

_____What methodology, instrumentation and statistical analyses were followed?

_____How does the present study fit in the context of existing literature?

_____Need for the study is established, with the likelihood of gaining relevant and significant results

_____Various theoretical positions and conceptual frameworks related to the problem are delineated and, if they affect the research, effectively denied.

_____Literature review is integrated, not a series of messages classified according to type or genre, but arranged according to questions, hypotheses and objectives cited in the problem section

_____Preference is given to primary sources

_____Preference is for works of the last decade; older works defended (except in historical studies)

_____Literature review includes a current awareness search of relevant journal articles of the past three years

_____Literature review is selective but sufficiently comprehensive for social research

_____Literature review is summarized

The Methodology

_____Research methodology is described (e.g., quasi-experimental, historical, case and field, correlation, causal-comparative, developmental, etc.

_____Research design is described

_____Relationship between statistical (null) hypotheses and research hypotheses (problem) established

_____Pilot study (if used) is described

_____Selection of subjects described

Instrumentation

_____Instrument is identified; defended for appropriateness, reliability, validity, etc.

_____Field procedures (case studies, observation, test of new instruments, etc.) specified

_____Data collection and recording process described

_____Data processing and analysis (statistics) described

_____Methodological assumptions and weaknesses cited

Title of Research

_____Problem is precisely identified, including specification of independent and dependent variables and target population

_____Title is clear and concise for indexing

_____Words are effectively arranged

General

_____Headings are used to identify logic and movement of proposal

_____Proposal contains suggested outline of the research.

_____Proposal contains a detailed time schedule for completion of each section

_____Proposal is written according to prescribed form and style

RESEARCH AND TECHNICAL WRITING

Topics Discussed in Chapter Eleven

- The Elusiveness Of The Writer
- Informal Versus Formal Writing
- Informal Writing
- Informal Emphasis
- Formal Writing
- Emphasis By Position And Proportion
- Diagnosing Writing Problems
- Examine - To Diagnose The Problem.
- Think - To Invent A Subject. (Invention)
- What Are You Going To Write?
- Why Are You Writing?
- What Do You Know About The Subject?
- Plan - To Arrange Ideas. (Disposition)
- Who Are The Readers?
- What Message Form Will You Use?
- Organize - To Present Facts. (Disposition)
- Write - To Express Yourself (ElocUtion)
- Revise - To Improve Style.
- Checklist For Writing And Revising The First Draft
- Revise - To Sharpen Sentences.

THEORY

INDUCTIVE LOGIC   DEDUCTIVE LOGIC

OBSERVATION

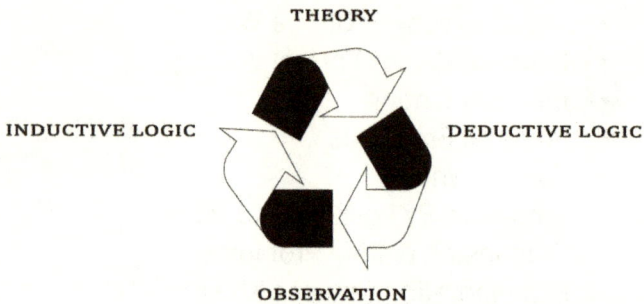

The Wheel of Research
Hunch + Assumptions + Assertions + Objectives + Lit Review +
Research Hypotheses + Statistical Hypotheses + Observations +
Conclusions + Generalizations

The Elusiveness of the Writer

A basic difficulty in understanding the written word is not the vagueness of the composition, but the elusiveness of the author. This could be the reason English/European students read authors rather than subjects. The more one knows the author of a text the easier it is to comprehend what is written. Knowing the orientation of the author and the background, philosophy, and purpose of the composition increases the understanding of the written document. For example, one understands a letter from a parent without much difficulty. An unsolicited letter from a stranger giving details about a subject of which one is not already somewhat informed could be perplexing. Yet, the correspondence of two professors is a common academic field interacting on a subject of mutual interest is readily comprehended.

Informal versus Formal Writing

The science of interpretation primarily relates to formal writing; however, the more the student knows about the process of composition the better the interpretation of written material will be. First, understand the word "composition" simply means with position. Every word, phrase, sentence, paragraph, section and chapter have a proper place in a composition. When all the components of writing are in the proper place or position, the process of understand the meaning and interpreting it to others is simplified. Take a look at both formal and informal writing, because some writers lapse for time to time into an informal mode. To adequately interpret what is written, this must taken into account.

Informal Writing

Informal writing is oral in nature. It is written for the **ear.** It is casual, simple, and spontaneous and requires little advance preparation. It is unassuming and unconventional. It is a script written to be read aloud or spoken--an oral script. Such writing is basically talking on paper. The emphasis is often lost unless one reads or mouths the words. It is conversational and designed for entertainment or popular reading, such as, novels or plays. TV scripts and TV and radio news are oral scripts, designed to be read aloud and heard with the ear. Emphasis depends on gestures, body language, or inflections of the voice. Informal writing is much easier than the formal, because the ear is a more sensitive sense. Hearing is an outgrowth of the touch, the most personal of senses. Consequently, hearing is a personalized way of touching at a distance. As such, hearing is the most social of the senses and has particular value in the context of informal writing. Informal writing is designed to touch the reader.

Informal Emphasis

The emphasis is often lost unless one reads or mouths the words. The writer uses underlining, bold type, italic, and other oral approaches to emphasize a word or phrase. This function by the writer is a substitute for speaking. Words and phrases are often mouthed or read aloud to determine the emphasis. The oral items above are then used to guide the reader.

Formal Writing

Formal writing is designed to be seen with the **eye.** It uses Standard English with prescribed or operational definitions. It is sometimes seen as awkward or stiff but always follows a conventional form and style based on the nature of the work and the audience. Since informal writing is talking on paper:

formal writing could be characterized as thinking on paper. The emphasis does not depend on the personality of the writer or the variations of voice or gesture. The emphasis is by position or proportion. It is structured to be read by the eye. Since the eye does not see when it moves, only when it stops to focus, punctuation becomes the road signs to stop the eye and enable it to see key words or phrases. By stopping the eye, words and phrases on both sides of the punctuation take on value. The key to formal writing is to keep in mind the words are to be viewed by the eye not heard by the ear.

Emphasis by Position and Proportion

Words and phrases are emphasized by the position in the sentence. Sentences are emphasized by their position in the paragraph and paragraphs are emphasized by their position in the article or chapter. The most emphatic position is at the end of the sentence. The second most emphatic position is at the beginning. The same holds true for the sentences in a paragraph and for paragraphs in a composition.

Value is awarded to paragraphs and chapters by both position and proportion. Those last and first have position for emphasis and should be shorter. The others must be emphasized by size and proportion. Those paragraphs and chapters, which fall between the first and the last, must be given size according to their position. Paragraphs lost in the middle of the composition or chapters stuck in the middle of a book, must be larger in size and more carefully written to get the attention of the reader and send the message of value for the contents. The closer these elements are to the first or last position the less the size. The closer to the middle the larger chapters must be to hold their value in the eye of the reader.

Here is the thinking is required. One thinks to invent a subject (title); then plans to arrange basic ideas (outline and paragraphs); next is the organization to present thoughts, facts, etc in sentences to support the topic sentence of the paragraph. Also, the paragraph must have a transition sentence to assist the reader in understand how one idea is connected to the next idea. Remember, a paragraph is one idea fully developed and so is a chapter. Chapters also need topic paragraphs and transition paragraphs for the reasons listed above.

Thinking produces the Book Title. **Planning** provides the outline of major ideas; thus Chapter Titles. **Organizing** assists with developing the outline of ideas for each paragraph. **Writing** provides a means for the author to express clearly. **Revision** requires patience and time, but it must be done in a timely manner, because a composition is a substitute for the author's words and must be clearly understood in terms of the author's qualifications, vocabulary, and intentions. **Time** is a factor in all revision. Surely one could do better if more time was giving, but there is always a deadline on a writing project. That is why most manuscripts are originally presented to a publisher in double spaced pages so that a Book Editor may made notations and changes between the lines. This must be honored to gain a dependable repetition and a timely book release. The best way to improve writing speed is to diagnose writing problems.

Diagnosing Writing Problems
The problems in writing create difficulties in adequate interpretation of what is written. Poor writers are not easily understood. The better the writer the easier it is to understand and interpret the meaning of written material. If the interpreter knows the normal pitfalls of writing, this fact can be observed when a writer lapses into poor style. Style

is simply the way a writer puts thoughts and words together. Poor writing is poor thinking. Good and logical writing is easily understood.

Examine - To Diagnose the Problem.
What is the writer's problem? An interesting person may be a dull, foggy writer because he does not consider the basic principles of clarity. Weakness must be recognized and corrected before a superior writing style can emerge. Most poor writers can find their faults in this list:

Wordiness. Readers are busy. They will not take time to figure out complicated vocabulary or plow through unnecessary words. Long, complicated words were constructed to think with, not to be used in basic communication. Some of this may be corrected by limiting pronouns, prepositions, possessives and contractions. Pronouns require an antecedent, prepositions require an object, possessives require an owner, and contractions require the reader to add words to the manuscript. All of this slows the reader and limits comprehension.

Poor Planning. It does not matter how fast or carefully you travel; if you are on the wrong road, the destination will not be reached. Planning time is as important as writing time. Poor planning triggers bad tone and poor organization. Before you write, think about what you are to say, plan how to say it, and organize your thoughts before starting the first draft.

Bad Organization. Work out an outline. Otherwise, there will be sentences and paragraphs going off in several directions. The lack of outline and structure is a serious writing weakness. Effective writing follows a logical and orderly progression of ideas. The main points are determined through planning. These are organized to present points most logically to the reader.

Weak Start. Beginning without an outline, poor writers go into a writing shell and strain for words and ideas. They lose the simple conversational quality essential to good writing and become overly concerned with form, grammar, and punctuation during the first draft. The mechanics can be improved during revision. Naturally, a good title and outline helps the writer to have a good beginning. Revision can and will make the writing more effective.

Sentences. Sloppy sentences are robbers of clarity. A good sentence conveys the exact meaning without embellishment. Sentences should be short unless the writer has mastered balance and other subordinating techniques. Beware of dangling, misplaced, and pyramiding modifiers, fuzzy statements, big words, passive voice, and topsy-turvy structure.

Transitions. Paragraphs are the major units of writing. Although the unadorned declarative sentence is a major accomplishment, the sentence must be properly placed into a paragraph. The sentence flow of long paragraphs must be smooth and the idea relationship clear. Remember that a paragraph develops one idea. These basic units of composition are linked with words or idea bridges, which help in the transition from one idea to the next. The reader must understand the relationship of ideas from paragraph to paragraph to understand when the writer has finished with one idea and is ready to move to another.

Words and Tone. Composition is a substitute for personal contact. The language, approach, and tone used should be polite and acceptable to the reader. Use short, familiar words; most people want information in simple form. A poor choice of words, a misplaced punctuation, or a twist of a phrase can cut deep into a reader's sensibilities, defeating the purpose of the composition. Watch for punctuation, style, spelling, and grammar; breakdowns here confuse the reader and distort meaning. When mechanics, word choice or tone go astray, so does emphasis and coherence. Revision is vital to correct these weaknesses. Reading the composition aloud helps. The ear picks up some things the eye cannot. The use of a good dictionary helps, too!

The interpreter's personal writing problems are also a hindrance to good interpretation. One who understands grammar, sentence structure, emphasis, etc. is better equipped to interpret written work. To improve personal ability to interpret written material, become a better writer. What is your basic writing problem? Perhaps you have more than one weakness. There is no advantage knowing about the weakness unless it is strengthened. These weak aspects of your journalistic skill should be strengthened by immediate action. Launch a self-improvement effort immediately. Read. Talk with others. Examine closely the work of good journalists. Work out a plan of self-improvement. Do some work in weak areas several days until you have developed a habit of self-improvement.

Do not worry about weaknesses. Strengthen them in revision. Some will always remain. The important thing is that you recognize and improve at your weakest points. Do not let weaknesses hinder you from writing. Rewriting can improve and make almost any manuscript acceptable. Of course, it may have to be reworked several times, but you can do it. Just get started!

Follow the TPO formula (think, plan, organize) before you write. This plan is based on some aspects of ancient rhetoric: invention, disposition, and elocution. Aristotle identified the first three steps in the rhetorical process as invention, arrangement, and style, but he added three aspects to bring credibility to the author: *ethos, pathos,* and *logos.* **Ethos** is a demonstration of what the author knows of the subject that influences the reader to accept the author's work as believable. **Pathos** is a construct relative to the author presenting a topic in a way that evokes strong emotions in the reader. **Logos** has to do with logical reasoning and objectivity in constructing a written document. Although these concepts are from the classical world of centuries ago, they remain good guideposts for writers.

Original writing is certainly a process of invention, but there are other elements of writing that are germane to this text; namely, *revising, rewriting, proofreading, re-wording,* and the list goes on and on, but the element of "time" is the limiting factor as the diagram below shows.   Always if there were more time one could do a better job, but reality is that there is never enough time to continue writing a composition over and over. One must do the best they can in the time that they have. This includes time for proofing and revising.

EW = T P O
WRRRRRR / RRR
T
Effective Writing = Think – Plan –Organize (TPO)
Write – read, revise, rewrite, revise, / revise
[time]

Think - To Invent A Subject. (Invention)

An inevitable act of writing is finding something to say. The what, why and who are the first hurdles. What are you going to write? Why are you writing? What do you know about the subject? What information will you need from other sources? Who is your reader?

Facts to support ideas and examples to explain ideas must come from somewhere, whether from your head, from research, or from a combination of both. Classical writers called this the process of **invention**. Actually, it means to find. The use of gray matter is essential to inventing a subject and finding something to say. There is no substitute for thinking.

What are you going to write?
A definite aim is essential to writing. Actually write down what the definite aim is before proceeding to choose a subject. The field must be narrowed to the specific aspect that can be covered in the proposed length of the composition.

Zeroing in on a specific subject assists the writer in several ways. It provides a starting place for reading and research. It guides in the developing of an outline to insure against sidetracks. Specifying a subject also helps identify the reader. This is a basic step in good journalism.

Why are you writing?
Decide! Writing to impress others with your knowledge will not produce a good composition. Write to be read. Keep in mind the three basic objectives of writing: (1) to inform, (2) to persuade, and (3) to interpret. Choose the proper one. Are you writing to arouse interest, influence attitudes, solve problems, implement progress, or evaluate results?

Should your goal be to inform state facts objectively and add to the reader's knowledge of the subject. To do this, a judgment must be made as to the present knowledge the reader may possess. When your goal is to persuade, attractive arguments and reasons should be used to stir the reader to decision. When you wish the reader to make up his own mind, write to interpret by analyzing facts and giving opinions.

Knowing why a composition is being presented helps the writer deal with specific issues. This improves the possibilities of good communication with the reader and lessens the chances of misinterpretation.

What do you know about the subject?

Why was this subject selected? Do you know a great deal about the field? Serious thinking is required at this point. To gain the reader's confidence, there must be no mistakes. The writer must have confidence in his knowledge of the subject. This is where he gets the authority to write, select details, use examples, and subject his work to the critical eye of process analysis.

What information will you need from other sources? Determining your limitations is the starting point. Research is the only way to compensate for prior lack of knowledge. Consider both primary sources and secondary sources. Talk and write to people. Read books, magazines, and papers. Interview, run tests, survey what others have done on the subject. Use your own library, the public library, and the libraries of educational institutions in the area.

Research until you have the facts to support your ideas. Think of examples to explain your ideas. Gather facts and list them in support of each idea or recommendation. No writing can be convincing without backup data.
Think of examples to explain your ideas. Examples perk up a paper and turn a dull subject into an interesting

composition. Facts show a writer's research; examples show his imagination.

Plan - To Arrange Ideas. (Disposition)
Next, based on your reading, list the ideas to be used in the composition. This list should include questions the reader may ask and ideas you think are important. From these questions, list in priority order which facts should be used to answer questions or support a point of view. This gives you the main ideas for the composition and provides a basis for organization of ideas into an outline for writing.

When ideas to be included in the composition are gathered, the next step is to find an effective order in which to present the information. Ancient writers called this "disposition." It was a process of ordering or arranging material. Disposition actually means, "to arrange." It is the business of deciding on the relevance and value of key facts, arranging them in groups and ordering those groups to achieve the desired effect on the readers.

Who are the readers?
Material cannot be properly arranged without knowledge of the readers. Who are they? What are their backgrounds? How much do they know about the subject? How will they use the information? How much time will they give to the composition? Are they likely to be receptive?

What message form will you use?
The message form has a great deal to do with how you write. The reader and his time combine to choose the message form. Will it be a letter, a report, a memorandum, an article, a telegram, or a book? Regardless of the form your message takes, the objective remains the same: communicate ideas concisely and clearly!

Organize - To Present Facts. (Disposition)

First, the specific purpose or thesis for the composition must be stated. The thesis has three elements: What, why, and who. What are you writing about? Why are you writing (inform, persuade, or interpret)? Who is your reader? The answer to these three questions provides a clear thesis. The order may be why, who, what in the final statement of the thesis.

Chapters, sections, paragraphs, thoughts must be presented in an organized format or the composition will not make sense. Readers react the same way to poorly organized writing as they would to the scenes of a drama being presented out of order. When thoughts are hard to follow, the reader gives up.

Trouble with organization is usually caused by a writer attempting to cover too much or because he cuts detail too fine. In such case the writer probably has limited himself to an outline of Roman numerals and capital letters. The outline is incomplete because he failed to plan the development and the presentation of detail. Some writers have too few headings; others have too many.

Without an adequate outline, there will most likely be poor transition between sections, paragraphs, and sentences. Important ideas will be buried in tangled thoughts and unbalanced structure.

Write - To Express Yourself. (Elocution)

When you have worked through invention and disposition (the THINK - PLAN - ORGANIZE sequence), you know what to say and where in the composition to say it. The remaining act of writing is simply to say it, to put your thoughts into the most telling prose you are capable of writing.

The Roman writers with oration in mind called this step **elocution** or to speak out plainly. This is essentially what is meant by the word "expression." The time has come to simply express yourself clearly. Put your ideas and thoughts into words. You know what you want to say and where you want to say it. There is nothing to worry about except expressing your thoughts on paper.

Remember you are writing on paper, not eternal bronze. Just put down the first sentence as it comes to mind. Complete the first paragraph. It may all be omitted in revision, but you cannot have a second paragraph until you have written the first. No one will publish it as it stands in the first draft. That is what revision is all about, but you have to have a draft to revise!

Most professional writers make a quick draft, often without giving it their best effort. They simply put their thoughts on paper knowing they can be sharpened up during revision. Even the most accomplished professional will never be satisfied with the effectiveness of a first draft composition. It is simply a first try. This means there are no good writers, but there are good re-writers.

The think-plan-organize sequence brings the writer nearly halfway through the writing process before the first draft. Do not worry about mechanics. Don't worry about form and structure. All this can be perfected later. Just put your thoughts on paper. Write to express your ideas. It is your idea, your pen, your words, and your draft. Do it your way! Let it sound like you. Permit your own style to come through.

Have you tried writing on index cards? What about colored paper? Perhaps ruled paper is best for you. Maybe you prefer to peck away on an old typewriter. Could grade school penmanship be your bag? Do you prefer a tape recorder, so you can write aloud? It is your choice. Do it your way. The way you feel comfortable. With adequate preparation and a comfortable place, you should be able to write without worry. The best place to start is always at the beginning, but you have done that. The think-plan-organize sequence was the starting place. Now you are not bound to any particular sequence. The outline is filled with good ideas. Look for a few key ideas. Then pick any key point on the outline and develop it, using the facts and examples gathered during research. Soon one sentence will follow another and you will be on your way to a first draft.

Of course, you were told not to worry, but you should worry a little. All writing needs a good beginning. You may start your composition at a key place in the outline, but eventually you must write the first sentence. The first paragraph(s) are important because they suggest to the reader what he can expect. Also, a good opening paragraph helps the writer produce a quick and more effective first draft by igniting a chain of ideas and setting a mood of confidence.

Revise - To Improve Style.
Style is the total effect of writing. It is achieved by the ideas, the organization, the paragraphs, the sentences, the words, all working together harmoniously to express the personality and thoughts of the writer. Style demonstrates the way a writer thinks and expresses their thoughts.

A negative reaction to a writer is a reader's expression of style recognition. From the reader's viewpoint, the composition is disliked or rejected because the reader does not appreciate the writer's style. When one dislikes a writer's style, he simply dislikes the way the writer thinks, expresses himself, his choice of words, sentence patterns, paragraph sequence, or transitions. Style then is usually recognized by negative reaction.

Revision gives the writer opportunity to practice this negative aspect of style recognition. Anything the writer dislikes must be changed and placed into acceptable form or style. Actually, a first draft is an inadequate expression and shows the writer certain areas where he lacks knowledge. When anyone considers a first draft a finished product, he does not understand the basic elements of writing.

The importance of revision cannot be overestimated. It is the difference between good writing and bad writing. Revision is not an unnecessary nuisance; it is an essential element of writing. No composition is complete without revision. Unrevised, it is only a draft, and a draft is not a completed product.

Revision requires the writer to switch roles. He must become an editor and a critic. It is more than flipping through the pages checking for sentence sense and obvious errors. The writer must stop being the creator and become the composition critic. This is the writer's role during revision.

Adequate revision gives the writer opportunity to judge the composition and prepare for reader reaction. More importantly, the writer has a chance to see his best ideas at work and has the privilege of revising and rearranging these ideas so they have the desired prominence and effect. Always leave space in the first draft for revision, leave room to work between the lines. This means you should double-space or triple-space the first draft. When space is not left for revision, a writer may be tempted to make another draft instead of revising the original. A writer is always better off using second draft time revising the first draft. Leaving space to revise is an important aspect of writing.

Do not revise immediately after writing the first draft. There needs to be a fallow period when the manuscript is forgotten. A composition can never be adequately criticized immediately after a draft is finished. The writer is too close to the problem to be objective. Remember the role change in revision. The writer is now an editor and a critic. Put the first draft away immediately. Do not tinker with it. Transfer your mind to another project and forget about the composition. Later you can be more objective.

It is best to write straightforward during the first draft, pausing as little as possible to make changes. It takes a special mindset to do this, but it is a more efficient way to express yourself on paper. No matter how well the draft is prepared, no one can anticipate all the traps and special problems that may occur in a manuscript. When the cooled-off draft is picked up for revision, the writer must brace himself for a shock. It may be worse than expected, especially at the beginning. This is no cause for discouragement. This is a part of the process!

The first few pages or paragraphs should be viewed carefully. The worst writing usually comes at the beginning, because the writer was trying to get the feel of the composition. The other place to look for special difficulties is just before the end. In a long work, careful revision is needed anywhere the writer was tired or bored. When the writer's energy slackens, he begins to write mechanically from notes. This cramps his style. Anything can happen. Careful revision is essential to effective composition. Revision must be orderly. It is not enough to read through the draft looking for something to change. There must be a plan. Try this one: (1) cut away the fat, (2) read aloud for content, (3) again read aloud for style, (4) watch for the common errors usually repeated. Each writer needs a self-improvement plan that includes a personal errors list. When the same mistake is repeated, it should go on an errors list. Use this list as a final step in revision. Try not to make the same mistake again.

Revision Plan

1. **Cut out the fat.** Wordiness is a constant difficulty with all writers. Vigor can be added to a style simply by cutting out all words unnecessary to effectively communicate the idea. Think of a telegram. Weigh each word, because each word affects style. Cut not only needless words, but unnecessary sentences and even poor paragraphs that do not contribute to the sense of the manuscript.

Go through the entire draft marking out unnecessary words. Use a wide felt pen to block out words, and phrases so completely that nothing remains to distract as the manuscript is read again. Look especially at the first few paragraphs or pages. The worst writing is probably there. Check the introduction. It may not make sense now. It may be necessary to cut a big slice off the beginning. The third paragraph on the second page may contain the kind of beginning that could capture reader interest and insure consideration of the composition. Keep this in mind as you cut out the fat!

Also, use pronouns, prepositions, and possessives only when the antecedent, object, or owner is clear. A pronoun must have a clear antecedent; a preposition requires an object; a possessive points to an owner. If these are not clear the reader will become confused or at least slowed in the reading process. Also, contractions require the reader to add words to the manuscript. Of course one cannot write in English without using pronouns, prepositions, and possessives, but they must be limited where the reader is not clear.

2. **Read aloud for content.** Reading the manuscript aloud is important, because the ear can detect errors the eye misses. Read aloud first for content. Read the trimmed-down draft aloud slowly listening to the sense of the composition. The fallow period and the cutting down should change the manuscript sufficiently to permit objectivity. Ask honest, critical questions; it is being read for content. Is the main point of each paragraph clear? Is the idea obvious? How is the transition? Has that been said before? Is it logical? Can it be simplified? Does it need more detail or an example?

When a question of clarity arises, begin to simplify at once. Confusion usually coincides with unnecessary complexity. When something seems unclear ask, "What do I really mean here?" The answer will simplify, clarify, and revise the problem passage.

Check the manuscript to see that each main point of the organizational outline is clearly expressed. Seek help from a friend. Ask that the composition be summarized by its major points. If a friend misses the main points, critics certainly will see the problem. Anytime the writer has to say, "What I meant was . . . ", something is missing or complicated about the manuscript.

3. **Read aloud again for style**. The manuscript has been repaired to insure sense and clarity, now it must be checked for rhythm. Listen for sentences with a sedative or washboard effect. Watch for verbal echoes and any unnecessary play on words. Cutting and revising may have caused difficulties of style that were not present before. Always observe punctuation as the manuscript is read aloud. Cutting may have left a series of similar short sentences that need attention. Listen for the complex as well as the over-simplified. Keep at least one ear open for

pretentious, pompous, formal style that does not sound natural.

Remember, style is the way a writer expresses personal thoughts and should be recognized as the writer's own vocabulary. Eliminate anything that does not sound natural. If it is not natural to the normal thinking of the writer, it probably will not be clear to the reader.

4. **Error list.** Finally, get out your error list. Remember, those same mistakes made each time - those little notes written on the side of each paper by a patient and helpful professor. There is no better check than the use of a personally accumulated error list.

Forewarned is forearmed! Following each typing, the manuscript must be copy-read for typographical errors. Blaming mistakes on a typewriter or typist can never erase mistakes on a manuscript. The writer is responsible regardless of who does the typing. Read the typist's work twice. Read once for sense and read again for word accuracy. The second time ignore the sense, check words for spelling and transposed letters. There are no typos on a manuscript that is complete; typographical mistakes are simply errors.

Always make a back-up disc and a hard copy of the manuscript. It is good insurance against the U.S. Mail's failure to deliver or the "wild blue yonder" called the INTERNET. In addition, reading a manuscript in the pre-published form is valuable to the writer. Copy-editing or publishers' proofreading will usually add to the writer's errors list.

A systematic approach to revision gives the writer control over the writing process. Augmented by a personal error list, this approach guarantees improvement. Application of these steps makes good sense. The world is overcrowded with poor writers. Ignore these techniques and you remain with the crowd. Use them and become the writer you want to be.

Checklist for writing and revising the first draft

1. Leave space for revision.
2. Wait at least 24 hours before beginning revision.
3. Expect the worst writing at the beginning or anytime you are interrupted.
4. First, cut out unnecessary words.
5. Next, read aloud for content.
6. Read aloud again for style.
7. Check draft against errors list.
8. Proofread at least twice: once for sense, once for individual words.
9. Accept responsibility for typographical errors.
10. Be proud of your composition!

Revise - To Sharpen Sentences.

The sentence has an effective place in the process of style. A workable technique to manage the writing of a sentence is almost nonexistent. A variety of subjective human variables account for this scarcity. Yet without these variables there would certainly be no eloquent style, for style springs from such individual sources. A few common sense guidelines do exist that can point to areas of improvement in the management of sentences.

The stylistic instinct must be awakened. This is the instinct, which comes with rhetorical gifts, which usually result from a great deal of reading. Vast reading experience seems to improve verbal aptitude, and naturalness grows that improves word selection and a sense of speech rhythm. This improves sentence control and emphasis. People who read regularly are unafraid to write. They feel at home in the world of words.

1. Observe the structure of the paragraph and the four paragraph essentials. A paragraph must (1) be founded on a topic sentence, (2) develop an evident progress of thought, (3) orient the reader to the thought process by use of connective and transitional signposts, and (4) make the reader aware of the place of the paragraph in the larger composition by some indication of function.

After the paragraph structure is clear, move to the next step. Read the material aloud, listening to the stages in the paragraph structure. Now, you are ready for the copying process, which should quickly give you insight into the writer's technique of his ideas in sentences and groups of sentences. On the surface, this process may seem unnecessary, but it is a sure way to develop an instinct for prose style and build naturalness into your writing.

2. Perhaps a review of the copying technique would be helpful. This is an instant reading experience that has a teaching value. Simply choose an author you like and a subject of interest. Select a paragraph or two, which you consider to be well written. The passage should be about twelve sentences in length.

3. The copying technique works this way. Read each sentence slowly aloud as you copy it by hand noting its pattern of emphasis and rhythm. Then copy the entire passage again trying to hear the sentences together in groups of two or three. By the completion of the exercise, you should have the rhythm and scheme of emphasis in your mind. The point of the whole process is learning to sense what sentences sound right. This can awaken the stylistic instinct.

Watch for rhythmic monotony in sentences. The elementary cause of this monotonous style is the sedative and the washboard effect in sentences. These often occur together in the work of inexperienced writers.

A series of sentences of the same length create the sedative effect on the reader. The monotonous sameness has the same effect as counting sheep. A bored reader, if he continues at all, will not understand. If two sentences of the same length appear together, do something to make the next one longer or shorter. Another way to break up the effect is to reduce one or two less important sentences to subordinate or dependent clauses with an important sentence serving as the main clause. These longer sentences will break the sameness.

A series of sentences beginning with their subjects creates a washboard effect and is a major cause of monotonous style. Inexperienced writers give their readers considerable discomfort with this weak sentence sequence. The cure for the washboard is the same as the sedative. As a rule, do not write more than two consecutive sentences beginning with subjects. Parallelism is to find elements that are mobile, such as adverbs, phrases, and subordinate clauses. Move these to the beginning of the sentence and break the washboard rhythm.

Simple and compound sentences are specialized sentences. A simple sentence has one independent clause. The compound sentence is two simple or two complex sentences joined by such words as *and, but, or, nor*, or *for*.

A strategy of sentence variety must consider the sentence used most often in writing: a complex sentence. Remember that the complex sentence has one independent clause and one or more subordinate or dependent clauses. The flexibility of the subordinate clause allows the complex sentence more variety of rhythm and emphasis than a simple or compound sentence. A skillful use of the sentence permits a reader to go through a series of sentences without sensing s sameness. You can spot complex sentences by the words that introduce the dependent thought or connect the two clauses, such as *since, although, after, as*, and *because*.

You may come across the terms *loose* and *periodic* sentences, which describe the emphasis of the major elements. If a sentence has the main idea at the beginning, it is a loose sentence. If the idea is at the end it is periodic. The periodic sentence is more emphatic than the loose sentence, but the strongest positions within the sentence remain at the beginning and the end. The best word order for a sentence is the natural one: subject - verb - object. It is the combination that will keep you out of trouble.

THEORY

INDUCTIVE LOGIC 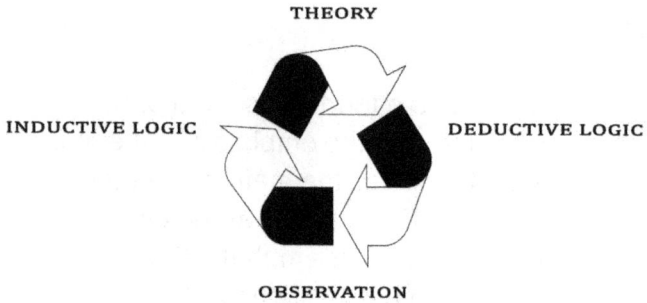 DEDUCTIVE LOGIC

OBSERVATION

The Wheel of Research
Hunch + Assumptions + Assertions + Objectives + Lit Review +
Research Hypotheses + Statistical Hypotheses + Observations +
Conclusions + Generalizations

CHAPTER TWELVE

RESEARCH AND CRITICAL THINKING

Topics Discussed in Chapter Twelve
- Critical Thinking uses Universal Principles
- Critical Thinking is a Persistent Process
- Critical Thinking is a Regimented Process
- Critical Thinking is Essential to Research
- Applying Critical Thinking
- A Learned Behavior
- Critical Thinking Skills
- Problem Solving is Critical Thinking
- Critical Thinking and Research
- Thinking About Research
- A Dissenting View on Research Methodology
- Skinner Summarized
- Value of an Informal Approach
- Serendipity* Findings

Learning without thought is labour lost.

— Confucius

Critical Thinking uses Universal Principles

Critical thinking is an important aspect of social research in order to carefully acquire and interpret data and reach a valid conclusion. Using universal thinking principles is both foundational and functional to any research project. One must remember that mistakes can happen due to researcher bias even when the methods of logical inquiry and reasoning are used. The identification and control of bias is part of the integrity of all research and critical thinking can assist in identifying the preconceived notions that damage the process.

Critical Thinking is a Persistent Process

As one thinks critically about problem solving, the process of thinking which normally expresses negative connotations must be put aside consciously and replaced with objectivity based on the tools of critical thinking. Such thinking requires knowledge of logical inquiry and the reasonable ability to apply these skills to the problem at hand. (Glaser, 1941) This is a persistent process that examines both the data and the process that produced the conclusions. Critical thinking in research requires the ability to recognize the formation opinions based on inconclusive data and comprehend patterns of relationships between propositions and data that may be confirmed or rejected. The use of appropriate thinking and language are essential to articulate this persistent process.

Critical Thinking is a Regimented Process

Critical thinking presupposes rigorous standards in thinking about problems and assumptions about antecedent influences that impact the present existence of an area of concern. Critical thinking in research is a disciplined and controlled process that assists with conceptualizing a problem, choosing operational methods, controlling the process, gathering reliable data and comparing variables in empirical grounded reasoning that leads to a valid conclusion. It involves assessment of research structures and essential elements in the research design and methodology. Critical thinking habitually uses competence in logic and proficiency in designing and conducting research to ethically guide the course of action in research. Combining research procedures and methods with the integrity of critical thinking brings both reliability and validity to the investigative protocol. The soundness of the process and the veracity of the conclusions are strengthened by critical thinking.

Critical Thinking is Essential to Research

Critical thinking is a crucial aspect of social scientific research; it is a process that assists with the control of bias. Flawed thinking inherent in some reports can cause bewilderment during the research process. Critical thinking avoids a one-dimensional view of difficult issues and produces a self-evaluation that exemplifies the principles illustrated in Plato's account of the trial and death of Socrates: namely, that the unexamined is not worthy of existence. In other words, each step in the research process deserves inspection and analysis to be worthy of inclusion. This is accomplished by using the necessary tools provided by critical thinking; the perceptions and principles that ensure adequate conceptualization and analysis of the constructs and operational terms of research. This vigorous action of imposing intellectual standards on the thinking process must be systematically cultivated and utilized in social research. When adequately used, critical thinking tools can improved confidence in the research process and assure the validity of research conclusions.

Applying Critical Thinking

Social scientific methods require the application of critical thinking in the research process. The close relationship to critical thinking tools and research methodology is easily observed. The skills needed in social research are matched in the process of critical thinking; or perhaps a better word is evaluative thinking. All academically prepared individuals developed critical evaluation techniques to reach their present level of achievement; however, these skills are not always applied to social research. Both critical thinking and scientific investigation require similar steps: problem identification and basic assumptions, hypothesis formulation and operational definitions, purpose/problem statements, scope and limitations, methodology and data analysis procedures, that included engaging the data that did not cluster around hypotheses or serendipitous findings that could enhance the contribution to scientific understanding of the research. This would include assertions of either theoretical contributions or practical solutions of specific problems.

A Learned Behavior

Critical thinking related to social research is a learned behavior and means correct thinking in the pursuit of relevant and reliable information about solving a particular problem. A critical thinker can normally ask appropriate questions, gather relevant facts, efficiently and creatively sort through data, reason objectively and decisively from this information, and processing the data accurately in a rigorous manner to arrive at valid conclusions. In other words, critical thinking involves critical investigation and as a learned behavior practice makes better, not necessarily perfect.

Critical Thinking Skills

Critical thinking skills are important to all students and researchers. The ability to think about a problem and express clearly and precisely those thoughts are crucial to valid reasoning. Clear thinking and evaluation of sources assists in filtering out the flawed thinking of others as one pursues the existing literature. Fact-based data becomes usable information for placing a problem in the context of the thinking of others. Constant evaluation is required when one attempts to interpret the words of an author. Information based on speculation or unconfirmed assumptions must be evaluated and utilized only to assist with the formulation of new assumptions about the problem but should not become supporting data for research. Perhaps the faulty information came from a lack of critical thinking by the original source. Alertness to such a possibility can produce vigilance and a critical analysis of all printed data. Such evaluation added to the information gathering process is the essence of critical thinking.

Problem Solving is Critical Thinking

Since problem solving is critical thinking, it is reasonable to conclude that problem solving research would have a connection to evaluative or critical thinking. Being able to distinguish between a statement of fact, opinion or an inference is an important contribution critical thinking makes to research. Critical thinking has a relationship to the scientific method, because the scientific process is the most powerful method to obtain relevant and reliable knowledge about a given subject. Knowing what can be supported empirically and legitimately derived from data is a crucial part of research. Critical thinking about a problem generates thinking in empirical and quantitative terms and these are key elements of critical thinking. It then becomes easy to understand why critical thinking has a popular manifestation in the scientific method.

Critical Thinking and Research

Critical thinking is related to research and problem solving because both require a careful and deliberate determination as to the acceptance or rejection of a conclusion and the degree of confidence with which the determination is accepted. Fundamental flaws and errors in research methodology may be detected through critical thinking. Research and critical thinking are connected at essential points: identify a problem and construct hypotheses, develop methodology, gather and analyze data, and make valid conclusions based on confirmed hypotheses. Critical thinking employs not only logic but broad academic research criteria and the appropriate theoretical constructs for understanding the problem, the gathered data, and the methods of analysis that produced the determination. The questions of reliability and validity must be answered.

Thinking About Research

Paraphrasing Glaser's influential study on critical thinking and making a present application to social research would identify thinking about research in three areas. There must be an objective attitude about the problems and subjects within the range of personal experience. Knowledge of research methods, data gathering and analysis skills, and systematic thinking about the process is required to do adequate social research. Finally, the application of appropriate and adequate proficiency with suitable methodology is essential to scientific investigation. (Glaser, 1941)

Two Philosophic Razors (Principles or Witty Sayings)

Ockham's Razor – is a principle attributed to the 14th century English logician and Franciscan friar, William of Ockham. The principle stated that the explanation of any phenomenon should make as few assumptions as possible, eliminating those that make no difference in the observable predictions of the explanatory hypothesis or theory. The principle is often expressed in Latin as the *lex parsimoniae* ("law of parsimony" or "law of succinctness"): *"entia non sunt multiplicanda praeter necessitatem"*, roughly translated as "entities must not be multiplied beyond necessity."

This is often paraphrased as "All other things being equal, the simplest solution is the best." In other words, when multiple competing theories are equal in other respects, the principle recommends selecting the theory that introduces the fewest assumptions and postulates the fewest entities. It is more often taken today as a heuristic maxim (rule of thumb) that advises economy, parsimony, or simplicity, especially in the scientific speculation. In this sense Ockham's Razor is useful in the area of social research.

Hanlon's Razor – is a corollary philosophic principle similar to Ockham's Razor that was simply stated, "Do not attribute to malice what can be explained by stupidity." A similar witty saying has been attributed to William James. Witty sayings have little place in social research, but it is always good to keep Hanlon's razor in mind.

A Dissenting View on Research Methodology
"The prerequisite of originality is the art of forgetting, at the proper moment, what we know." – Arthur Koestler

In research it is always important to consider variant material. This dissenting view on research is planned and systematic, ideally, to deduce hypotheses from theory and put them to the test or to set objectives, arrange procedures, and intentionally achieve a planned mission.

B. F. Skinner, the psychologist, whose impact on education is only beginning to be felt through such innovations as the modern teaching machine, operant conditioning, and behavior modification, vigorously dissents from any formal

view of "the scientific method." In remarks from his book
Cumulative Record (1959), Skinner discussed what he
called the informal "principles of scientific practice." He
noted:

1. When you run onto something interesting, drop
   everything else and study it.
2. Some ways of doing research are easier than others are.
   [In searching for an easier way to do something, one
   may develop more efficient techniques and also discover
   unforeseen phenomena.]
3. Some people are lucky. [New ideas and discoveries often
   are accidental.]
4. Apparatuses sometimes break down. [And one stumbles
   upon unexpected and fruitful consequences.]

Skinner Summarized
   This account of my scientific behavior is as exact in letter
   and spirit as I can now make it. The notes, data, and
   publications which I have examined do not show that I
   ever behaved in the manner of Man Thinking as described
   by John Stuart Mill or John Dewey or in reconstruction
   of scientific behavior by other philosophers of science. I
   never faced a problem, which was more than the eternal
   problem of finding order. I never attacked a problem by
   constructing a hypothesis. I never deduced Theorems or
   submitted them to Experimental Check. So far as I can
   see, I had no preconceived model of behavior—certainly
   not a physiological or mentalistic one and, I believe not a
   conceptual one.   — B.F. Skinner

Value of an Informal Approach
Another strong argument for the value of an informal
approach to scientific research is presented by Arthur
Koestler in his scholarly work The Act of Creation (1967) a

comprehensive documentation of the thesis that discovery, invention, and originality whether in science, technology, or the arts is remarkably unsystematic, unforeseeable, and "accidental."

Serendipity* Findings
*Serendipity – according to Noah Webster is a word coined by Horace Walpole about 1754 after his tale The Three Prince of Serendip (i.e. Ceylon), who made such unexpected discoveries. The meaning became "An apparent aptitude for making fortunate discoveries accidentally."

The informal approach to research is reminiscent of the fairy tale and the Princes of Island of Serendip who periodically would venture to the mainland in search of something. While they never accomplished their intended task, they always returned with new discoveries or experiences more marvelous than their original objective. For this reason, the term "serendipity" came to mean "the finding of valuable or fortunate discoveries accidentally."

APPENDICES

Appendix One

Problem Solving Techniques

Topics Discussed in Appendix One
- Problem To Find
- Problem To Confirm
- From Example To Pattern
- Try This Yourself
- The Example Becomes A Pattern

PROBLEM SOLVING IS THE GOAL OF SOCIAL RESEARCH.

Below are two areas where one may practice problem solving skills. It is good to remember that all social research is directed toward the solution of a problem.

A Problem Solving Technique Applied to
The Teaching and Learning Process

What is a problem? A problem exists when one searches consciously for some action appropriate to attain a clearly conceived, but not immediately attainable objective. Solving a problem means finding such action. Problem solving is finding a way: a way out of a difficulty, a way around an obstacle.

Solving problems is a specific achievement of intelligence and can be regarded as the most characteristically human activity. The conscious thinking is concerned with problems. When the mind is not engaged in mere daydreaming, it is directed toward certain objectives and seeks ways and means of reaching these goals.

Problem solving is a practical art: one can learn it only by imitation and practice. There are no magic keys that open all the doors and solve all the problems, but life gives good examples for imitation and many opportunities for practice. Those who want to become a problem solver, have to solve problems.

Four principal characteristics of a problem solver have been determined by studies on the subject:

- **Problem Sensitivity**
  Recognizing that a problem exists.

- **Idea Fluency**
  Compiling a large number of alternative solutions.

- **Originality**
  Using original variations to meet existing conditions.

- **Creative Flexibility**
  Considering a wide variety of approaches.

Analyze the Problem
The existing difficulty must be analyzed and classified. Take time to study the facts involved. In every problem there is (1) an unknown, (2) something known or given (Data), and (3) a condition which specifies how the unknown is linked to the data. The condition is an essential part of the problem.

What kind of problem is this? If a problem can be classified and recognized as a type, progress has been made toward finding a solution. The method learned to solve this problem must now be recalled. A good classification should suggest the type of problem and the type of solution.

There are two general types of problems: one is to complete what is missing, the second is to support an assumption. There are "to find" and "to confirm" problems. The aim of a problem "to find" is to (construct, produce, obtain, or identify) whether a certain assertion is true or false, to affirm or reject it. When one asks, "How can I find a way?" a problem to find is presented. Yet, when one asks, "Did he find the way?" a problem to confirm is presented.

Problems to find may be compared to the goals of the teaching process:

        1)   stimulate interest
        2)   arouse a spirit of inquiry
        3)   get the learner to work

Problems to confirm may be compared to the learning process:

        1)   memorization
        2)   understanding
        3)   expressing the thought
        4)   giving evidence
        5)   applying knowledge to life

Problem to Find
The goal of a problem to find is to locate a certain object, the unknown of the problem, satisfying the condition that relates the unknown to the data. A problem clearly stated must specify the category to which the unknown belongs and the condition the unknown is to satisfy. Is the unknown a teaching method, a word to reach the heart, or an action to deal with a discipline problem?

Taken in a strict sense, the problem demands one find (identify, produce, construct, identify, list, characterize) all solutions. Taken in a less strict sense, the problem may not have just one (any one) solution. For practical problems dealing with something similar to the teaching and learning process, the "strict sense" would make little sense.
A problem that is not understood cannot be solved. To understand a problem one must know the principal parts: the unknown, the facts or data, and the condition relating these facts to the unknown. Therefore, it is advisable in problem solving to pay close attention to the principal parts of the problem.

The basic problem of the teacher is to link the facts of the lesson to the lives of the learners. The teacher must find a way to excite and direct the self-activity of learners, and as a rule, tell him nothing they can find out for themselves. Teaching and learning is a cooperative venture.

Problem to Confirm:
The principle parts of a problem to confirm are called the hypothesis and the conclusion from the tentative assumptions. To support the proposition one must discover a binding link between the principal parts, the hypothesis and the conclusion. The hypothesis should suggest the conclusion.

What are the facts in the data? What does the learner understand? Can the learner personally express the thought in common language? Can evidence be given for the facts learned? Have the facts been applied to the life of the learner? These are problems to confirm and must be supported by evidence produced by the learner. Whoever takes these questions seriously has presented a "problem to confirm."

Where there was no learning there was no teaching. By applying this test, the teacher confirms whether or not the instruction was effective. When the teaching process was successfully completed, the facts of the lesson should be reproduced in the life of the learner.

From Example to Pattern
Each problem solved provides guidelines that serve afterwards to solve other problems. In each problem solving experience, there are features that may be useful in handling future problems. A personally obtained solution or one read or heard about, but followed with real interest and insight, may become a pattern, a model that could be used with advantage in solving similar problems. Problem solvers could

learn skills by becoming familiar with two basic patterns: [The teaching/ learning process is used only as examples.]

Example #1: Construct a triangle being given its three sides.

Try This Yourself
With equal line segments A, B, and C place A between points B and C, then draw two circles, using center C and radius B, the other with center B and radius C; let point A be one of their two points of intersection. The ABC is the desired triangle. The unknown was a point, the third vertex of the triangle.

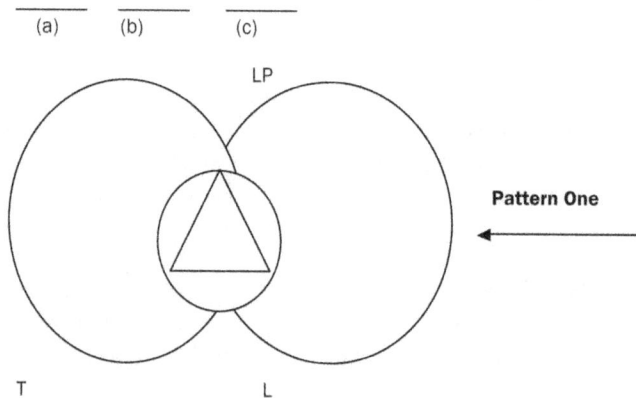

More Information Is Gained As One Progresses

By laying down the segment A, one locates two vertices of the required triangle, B and C: just one more remains to be found. In fact, laying down the segment transformed the proposed problem into another problem equivalent to, but different from, the original problem. In the new problem, the unknown is a point, the third vertex of the triangle, the data are two points B and C and two segments and the condition requires that the desired point be at the distance B from pint C and at the distance C from point B.

Example #2: Circumscribe a circle about a given triangle.

A Little Harder But Try This One.
The data of the problem are three points A, B, and C. The condition concists of the equality of three distances. First, one must find the center of the triangle, point D, and use this point to construct a circle about the triangle.

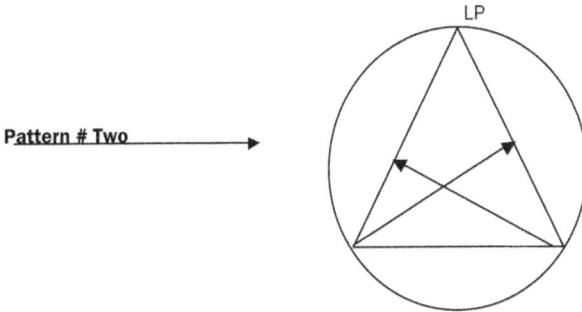

LP

Pattern # Two ⟶

To solve the problem, divide it into two parts. Find the center of the triangle and then construct the circle. The center is the point where the perpendicular bisector of segments AB and AC intersects. Use this point to construct the circle.

The Example Becomes a Pattern
Pattern #1:   Construct the triangle of the teaching process being given its three goals.

The goals are (a) stimulate interest, (b) arouse a spirit of inquiry, and (c) get the learner involved. First step is to (a) make a connection between the teacher (T) and learner (L) by stimulating thinking. Next (b) bring the learner within the circle of the teacher's influence and knowledge by arousing the spirit of inquiry. Finally, (c) the teacher responds to the learner's inquiry by getting the learner to work for information. Then the learning process (LP) begins: the teaching process has initiated the learning process. This gives the third point of the triangle (T L LP).

Pattern #2:   Circumscribe a circle of facts about the learner being given a completed teaching triangle T L LP

This involves completing the learning process: (1) memorization, (2) understanding, (3) expressing the thought, (4) giving evidence, and (5) applying knowledge. The completed triangle of the teaching process must be used to continue the learning process until the heart and mind of the learner is surrounded with the facts of the lesson.

The three goals of the teaching process form the given triangle. Stimulating the learner's interest in the lesson and creating an awareness of the facts (memorization). The learner seeks understanding when the spirit of inquiry is aroused. When the teacher gets the learner involved, the learner is able to express the thought of the lesson and the teacher starts the process of testing the learning.

The first three steps have been taken, but the facts of the lesson must reach the learner's heart and mind and influence both attitude and behavior. By bisecting the sides, the center of the triangle is located. It works this way: memorization and understanding lead to expressing the thought, and memorization and expressing the thought supports the learner's understanding of the lesson. Where these two areas overlap is the core of the learning problem. The teacher must now use this understanding to solve the learning problem and influence both the attitude and behavior of the learner. This is the completed circle in the teaching and learning process.

A Problem-Solving Program For Defining
A Problem And Planning Action

This problem-solving program is designed to assist one in analyzing a problem in organization, management, or human relations-- any problem which involves people working or living together. In this exercise one learns a new way of thinking about change. This program can assist one in analyzing something that needs to be changed and learn how to make the change without confusion and discomfort.

This is a programmed workbook. That is, it is presented in a series of separate steps of "frames," each of which contains a complete and separate idea, questions, or instruction.

Frame One: Identifying the problem.
The first step in the process of analysis is to identify the problem on which you wish to work. Here are some guidelines for selecting the problem:

--Is it something you really care about? Select a problem on which you want to work rather than a big abstract social issue in which everyone should be interested. A teacher, for example, may be concerned about improving the quality of education, and might select "maintaining better discipline in the classroom" as a problem.

--<u>Is it something with personal involvement?</u> Be certain the problem is intimately personal. A teacher might be concerned about discrimination in the community, but the problem might be "a personal attitude toward individuals of another race or culture."

--<u>Can you personally influence the situation?</u> Personal influence is normally wider than one realizes, but there are limits.

Identify the problem below:

Frame Two: Restating the problem.
Most problem statements can be rephrased so they describe two things:

    a) **situation as it is now (the real)**
    b) **situation as one desires it to be (the ideal)**

Briefly restate the problem below, indicating the desired direction of the change. Example:

Poor: I would like to reduce the amount of arguing in my family.
Better: Reducing family arguments.

Poor: My ability to delegate responsibly.
Better: Increasing my ability to delegate.

Restate the problem in terms of directional change below:

Frame Three: Restraining and Driving Forces
Most problem situations can be understood in terms of the forces which push toward improvement and the forces which resist improvement and keep the difficulty as a problem. For example, some forces reduce family arguments; such as, the love and affection between family members, room enough for privacy, etc. These could be called Driving Forces. Other forces tend to increase family arguments; such as, lack of closet space, not enough bathrooms, the generation gap, etc. These are restraining forces.

What are the driving and restraining forces affecting the situation? Think about these in broad terms. Include personality factors, physical resources, and restraints, feeling social pressures. List anything that comes to mind, without being critical or selective. Weed out the irrelevant items later.

Restraining Forces   **Driving Forces**

Frame Four: Identifying Specific Forces
Now review the two lists, and underline those forces which seem to be
the most important, and could be affected constructively. Depending on
the problem, there may be one specific force which stands out, or there
may be two or three driving forces and two or three restraining forces
which are particularly important.

Frame Five: Reducing the Influence of Restraining Forces
Now, for each restraining force underlined, list some possible action
steps which might be able to reduce the effect of the force or to
eliminate it completely.

Brainstorm. List as many action steps as possible, without worrying
about how effective or practical they may be. You will later have a
chance to decide which are the most appropriate.

List Action Steps

Restraining Force
A:_____
Possible action steps to reduce this force:

Restraining Force
B:_____
Possible action steps to reduce this force.

Restraining Force
C:_____
Possible action steps to reduce this force.

Frame Six: Increasing the Influence of Driving Forces
Now, do the same with each driving force underlined. List all the action
steps which come to mind which would increase the effect of each
driving force.

Driving Force
A:_____
Possible action steps to reduce this force.

Driving Force
B:_____
Possible action steps to reduce this force.

Driving Force
C:_____
Possible action steps to reduce this force.

Frame Seven: Beginning at the Stress Points.
The place to begin to solve the problem is at those points where
some stress and strain exist. Increased stress may lead to increased
dissatisfaction which may be a motivation for change.

Sometimes an attempt to increase a driving force results only in a
parallel increase in the opposing force. Consider whether the change
would be managed more easily by reducing a resisting force. Review the
listed action steps and underline those which seem promising.

Frame Eight: Listing Action Steps and Available Resources
List the steps underlined. Then for each action step list the materials,
people, and other resources which are available for carrying out the
action.

      Action Steps   **Resources Available**

Frame Nine: Review for Plan of Action
Now review the list of action steps and resources in the previous frame, and think about how they might each fit into a comprehensive action plan. Eliminate those items which do not seem to fit into the overall plan, add any new steps and resources which will round out the plan, and think about a possible sequence of action.

Frame Ten: Devising an Evaluation Plan
This step in the problem-solving process is for you to plan away of evaluation the effectiveness of your action program as it is implemented. Think about this now, and list the evaluation procedures to be used.

Frame Eleven: Prepare a Problem-Solving Proposal on the Problem

Now a plan of action is in place to deal with a problem situation. The next step is to prepare a Problem-Solving proposal to present to the faculty for evaluation before the plan is implemented. The length, form, and style of this proposal is to be determined by the nature of the problem and the student. Faculty evaluation will be two-fold:

1. A member of the faculty will evaluate the plan and discuss the implementation.

2. The student will implement the plan and present a report and evaluation to the faculty member for credit.

This concept originally designed by Saul Eisen and Cyril R. Miller for a training laboratory for NTL Institute was adapted with permission by Hollis L. Green for Oxford Graduate School. @ 1983.

Appendix Two

Research and the Internet

Topics discussed in this section

- Introduction
- Planning Your Research Project
- Finding Resources
- Directories
- Search Engines
- What Are Search Engines And How
  Do They Work
- Planning A Search On The Internet
- Making The Most Of Google
- Tips On Conducting Internet Searches
- Google Scholar
- Searching An Online Book Or Journal
- Understanding A Google Scholar Search Result
- Explanation Of Links
- Searching Google Scholar
- Windows Live Academic
- Using The Wikipedia
- Evaluating Internet Resources
- How Old Is The Material?
- Who Wrote The Information?
- Why Is This Material Here?
- Can I Do A Cross Check?
- Evaluating Internet Resources
- Plagiarism & Referencing Electronic Resources
- What Is Referencing?
- Why Reference?

- Apa Citation Of Electronic Resources

Publisher's Note:

The Internet and Information Technology is expanding; it appears that nothing stays the same. New ways of accessing information and the knowledge index are opened almost daily. New web sites appear hourly on the Internet. All who search the Internet for data should be aware of these changes. Also, students and new researchers should be aware that all the information on the Internet is not useful or worthy of consideration. Mostly, the problem is the illusiveness of the author who presented the data.

The Internet is most helpful in finding public knowledge information, government and corporate documents, and biographical sources, but the original documents and/or published works are the true source. It should be noted that most of the information on the Internet has not been adequately refereed. With the current errors in commonly used textbooks that are closely edited, one should be aware that most of the Internet information does not have a scholarly filter. In fact, the Internet has become more of a social network than a source of academic data. Those who access and use data from the Internet must be both cautious and discerning in the Internet sites (data bases) accessed to retrieve scholarly information. Finally, one rule of statistics could be helpful: "All data contains error."

Apprendix Two

Research and the Internet

Appreciation is extended to Joshua D. Reichard, DPhil, for his work on this **Guide to Internet Research.**

Introduction

The Internet is the world's largest computer network. Through this network, users from anywhere in the world can connect and share information. The 'World Wide Web' is the most popular part of the Internet; it is the multimedia component and is navigated by clicking on hyperlinks or hypertext. Although the terms are often used interchangeably, when we talk about 'searching the Internet' for the most part what we are really referring to is using the mainstream 'Web' component of the Internet to locate and share information.

There are no formal controls restricting what is made available via the Internet. This means that it is important to consider carefully how you are going to search for information and which tools you are going to use. The information provided here is intended to assist you conducting effective searches relevant to your information needs, whatever they may be.

The Internet is an exhaustive resource for researchers, but like every other type of research tool, new skills must be learned and decisions must be made about how and when to use it. This is where your research planning will be important.

Planning Your Research Project

Some of the points you need to consider are:

• What is your research problem? (i.e. what is it you are looking for?).

• What kind of information do you need?

• What tools are you going to use to find this information? (i.e. the Internet, the library)

• How are you going to find this information? (i.e. if you are using Internet search engines, what keywords are you going to use?)

- How much time are you going to spend researching? (i.e. you could go on forever finding information, but you shouldn't. What would be a reasonable time frame for finding the information you need?)

- How are you going to evaluate your information? (i.e. when you are finished your research it is always useful to reflect on the process. What things worked, what things didn't, what would you do next time?).

Finding Resources
Finding resources on the Internet can be like looking for a needle in a haystack. There are billions of documents on the Internet, published by specialists, scientists, teachers and students. Some of them will be useful for your research project. There is an array of research tools you can use to locate the information you need.

Directories
One of the easiest and safest methods of researching for relevant resources is by using Directories that have already been vetted by other organizations. Directories are collections of resources organized into categories. Sometimes the directory will focus on one subject area; others may collect and organize resources in a number of areas. Internet Directories include:

- The World Wide Web Virtual Library
http://www.vlib.org

- The Internet Public Library
http://www.ipl.org

- DMoz – The Open Internet Directory Project
http://www.dmoz.org

- Google Directory
http://www.google.com/dirhp

- InfoMine – University of California
http://infomine.ucr.edu/

- AcademicInfo Network
http://www.academicinfo.net/

- 	The Librarian's Internet Index
http://lii.org/

- 	Resource Discovery Network (UK)
http://www.rdn.ac.uk/

A useful directory for serious researchers is The Argus Clearinghouse (http://www.clearinghouse.net/) which "provides a central access point for value-added topical guides which identify, describe, and evaluate Internet-based information resources." While The Argus Clearinghouse is no longer being actively maintained, it remains a useful resource. An example of the categories of Sociology in the DMOZ Directory

Search engines

What are search engines and how do they work?

The Internet is not like your graduate school or public library, with its shelves of well organized books. Simply browsing the internet is unlikely to find you the information you need, so in order to find the resources you want for your research project, you will need, at some point, to use a search engine.

Behind the scenes search engines compile databases of web pages which allow users to search the internet for specific resources, by doing what is termed a *keyword search*. When a user types in a search request such as "egypt", the search engine already knows where all the pages including "egypt" are located.

The search engines use "bots" or "spiders" which prowl the internet collecting pages, but depending on the search engine, the databases of pages may be more or less up-to-date.

Planning a search on the Internet
When you are using a search engine it's important to clearly define your keywords. You need to be specific rather than general, because there is so much information available; a general search may return you hundreds of thousands of hits. In order to avoid being overloaded with information, think carefully about what you are searching for.

For example, a search on Alta Vista for the word "religion" returned 536,000,000 hits! But a search for "sociology of religion" returned only 45,000,000 hits, still a large number, but much more useful.

Before you start your search think about what you are looking for and do some preliminary work with a pen and paper. Think of all the possible terms you might use for your subject. Think of any differences there may be in terminology from country to country

Each search engine uses a slightly different language to help you with your searches, so it is worth the time it takes to read the search guides each engine provides. There are many different search engines, but unquestionably, Google ranks as the most popular and most user friendly: http://www.google.com.

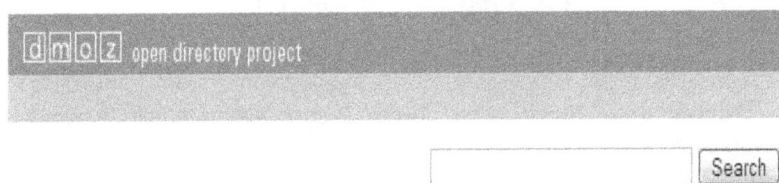

dmoz open directory project

[Search]

Top: Science: Social Sciences: Sociology *(1,090)*

- Academic Departments *(795)*
- Academic Papers *(17)*
- Associations *(48)*
- Ethnomethodology *(40)*
- Journals *(81)*
- Methodology@ *(212)*
- Sociologists *(41)*

- Cyberculture@ *(14)*
- Rural Sociology *(23)*
- Sociology of Religion@ *(15)*
- Sociology of Soccer@ *(8)*

See also:

- Science: Social Sciences: Anthropology *(943)*
- Science: Social Sciences: Criminology *(39)*
- Science: Social Sciences: Demography and Population Studies *(108)*

Figure 1.1 - A Keyword Search

Making the most of Google
There are some tactics that will assist you in utilizing Google to its maximum advantage. A Google search can be much more complex than a series of keywords. To start, if you are search for a specific phrase, be sure to enclose your phrase in double quotes. This ensures that the entire phrase is searched and not each individual word:

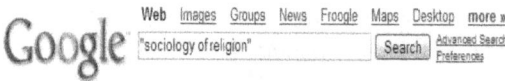

Figure 1.2 – An encased Google Search

Google will also search for pages for you by relevance to your keywords. To ensure that the exact phrase or word is specifically included in your search, add a plus sign (+) in front of your search term:

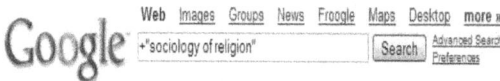

Figure 1.3 – A required phrase Google Search

To ensure that pages with specific keywords are not included in your search, add a minus sign (-) in front of the keywords:

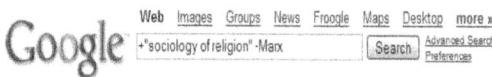

Figure 1.4 – An excluded phrase Google Search

These strategies will help you to narrow your search to the information most relevant to your research.

Tips on conducting Internet Searches
Read the directions at each search site. The technique for formulating a search depends on the search engine you are using. There is a wide variety of options available among the different search engines.

If you have a multi-term search, be sure to determine which type of Boolean logic you should use. For example, a search about the

relationship between latitude and temperature can be formulated as:
+ latitude   + temperature on many Web search engines in order for
AND logic to apply.

Include synonyms or alternate spellings in your search statements and
connect these terms with OR logic. Check your spelling. Take advantage
of capitalization if the search engine is case sensitive. If your results are
not satisfactory, repeat the search using alternative terms. Try different
sources to diversify your results. Sources can include other search
engines and large directories. Experiment with different search engines.
No two search engines work from the same index. Try search engines
which allow you to search multiple search engines simultaneously. Be
aware that you will lose access to advanced query options since not all
engines offer them.

If you have too many results, or results that are not relevant: Field
search. Add concept words to your original search. Use vocabulary that
is specific to your topic; avoid words with large concepts unless you
intend to field search. Link appropriate terms with the Boolean and (+)
so that each term is required to appear in the record. While many search
engines do not require this, it doesn't hurt to be on the safe side. Use
term proximity operators if they are available to locate documents in
which your terms are close together. Exalead is one of the few engines
nowadays that offers this.

If one of your search terms is a phrase, be sure to enclose it within
quotations, i.e., "global warming." Use the Boolean not to keep out
records containing terms you don't want. If you have too few results:
Drop off the least important concept(s) to broaden your subject. Use
more general vocabulary. Add alternate terms or spellings for individual
concepts and connect with the Boolean OR. Try the option available on
some engines to find similar or related documents to one or more of
your relevant hits.

Google Scholar is a relatively new search tool provided by Google that
allows a specific search of WebPages with academic citations. http://
scholar.google.com

Figure 2.1 – A Google Scholar Search

A search with Google Scholar will return much different results than the standard Google search:

Scholar

[BOOK] The **Sociology of Religion**
M Weber - 1993 - books.google.com
Page 1. THE SOCIOLOGY OF RELIGION Th j On e IJlliilllliiJllllliill RQQD-XWO- 5OY
Page 2. THE SOCIOLOGY OF RELIGION Max Weber Introduction by Talcott Parsons ...
Cited by 197 - Web Search - Library Search

The Contribution of Religion to Volunteer Work - group of 2 »
J Wilson - Sociology of Religion, 1995 - questia.com
... Contributors: Thomas Janoski - author, John Wilson - author. Journal Title: **Sociology
of Religion**. Volume: 56. Issue: 2. Publication Year: 1995. ...
Cited by 38 - Web Search - BL Direct

[BOOK] Why conservative churches are growing: a study in **sociology of religion**
DM Kelley - 1972 - Harper & Row
Cited by 79 - Web Search - Library Search

Recent Developments and Current Controversies in the **Sociology of Religion**. - group of 5 »
DE Sherkat, CG Ellison - Annual Review of Sociology, 1999 - questia.com
Recent Developments and Current Controversies in the **Sociology of Religion**>> <script
language="JavaScript" type="text/JavaScript"> function init ...
Cited by 55 - Web Search - BL Direct

Figure 2.2 – Results of a Google Scholar Search

Google Books
The first result of the Google Scholar search above lists the book "The Sociology of Religion" by Max Weber. When you see the [BOOK] tag next to a search result, this means that the book is available for searching and browsing online.

Searching an Online Book or Journal
There are immense advantages to searching through online versions of scholarly books and journals. Unlike the traditional Index of a book (which some may or may not have), an online search allows you to scan the entire text for specific key words in the same way you search the Internet. This allows you to target your research without wasting hours of time paging through irrelevant information.

In the example below, the search box for the Google book is on the lower-right side of the screen.

Understanding a Google Scholar Search Result
Each Google Scholar search result represents a body of scholarly work. This may include one or more related articles, or even multiple versions of one article. For example, a search result may consist of a group of articles including a preprint, a conference article, a journal article, and

an anthology article, all of which are associated with a single research effort. Grouping these articles allows us to more accurately measure the impact of research and to better present the different research efforts in an area.

Each search result contains bibliographic information, such as the title, author names, and source of publication. One set of bibliographic data is associated with the entire group of related articles and is our best estimate at the representative article for the group. This bibliographic data is based on information from the articles in the group, as well as on citations to these articles from other scholarly works.

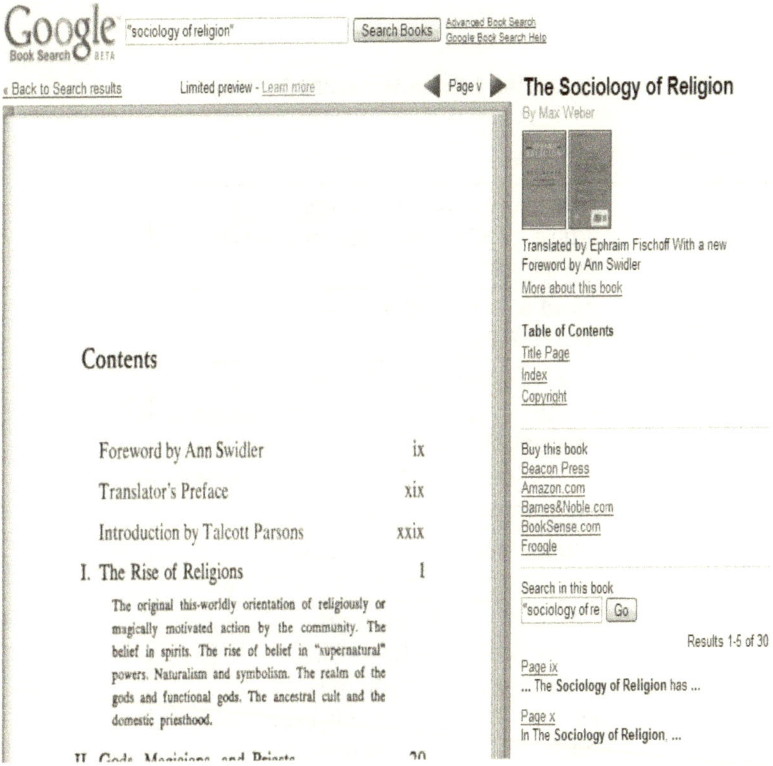

Figure 2.3 – A Google Book Search A summary of the Google Scholar features is included on the next page.

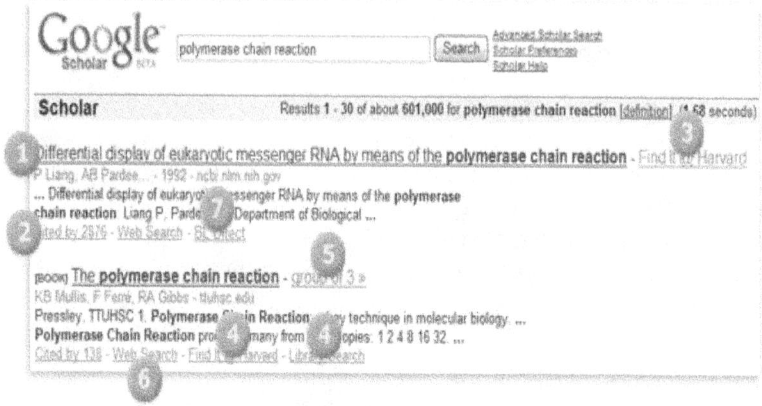

Figure 2.4 – A Google Scholar Search Result

Explanation of Links

	Title – Links to the abstract of the article, or when available on the web, the complete article.
	Cited By – Identifies other papers that have cited articles in the group.
	Library Links (online) – Locates an electronic version of the work through your affiliated library's resources. These links appear automatically if you're on the campus of a college or university.
	Library Links (offline) – Locates libraries which have a physical copy of the work.
	Group of – Finds other articles included in this group of scholarly works, possibly preliminary, which you may be able to access. Examples include preprints, abstracts, conference papers or other adaptations.
	Web Search – Searches for information about this work on Google.
	BL Direct – Purchase the full text of the article through the British Library. Google receives no compensation from this service.

Searching Google Scholar

How do I search by author? Enter the author's name in quotations: "d knuth". To increase the number of results, use initials rather than the full first name.

If you're finding too many papers which mention the author, you can use the "author:" operator to search for specific authors. For example, you can try [author:knuth], [author:"d knuth"], or [author:"donald e knuth"].

How do I search by title? Put the paper's title in quotations: "A History of the China Sea." Google Scholar will automatically find the paper as well as other papers which mention it.

How do I find recent research on a particular topic? Just click on "Recent articles" on the right side of any results page, and your results will be re-sorted to help you find newer research more quickly. The new ordering considers factors like the prominence of the author's and journal's previous papers, as well as the full text of each article and how often it has been cited.

How do I search for papers in specific publications? Within the Advanced Search page, you can specify keywords which must appear in both the article and the publication name. See Google's Advanced Search Tips for more information.

How do I search by category? From the Advanced Search page of Google Scholar, you can search for scholarly literature within seven broad areas of research. Simply check the boxes for the subject areas you're interested in searching.

☐ Biology, Life Sciences, and Environmental Science
☐ Business, Administration, Finance, and Economics
☐ Chemistry and Materials Science
☐ Engineering, Computer Science, and Mathematics
☐ Medicine, Pharmacology, and Veterinary Science
☐ Physics, Astronomy, and Planetary Science
☐ Social Sciences, Arts, and Humanities

Figure 2.5 – Google Scholar Categories

Windows Live Academic

Windows Live Academic lets you search academic journals and content for article titles, author names, article abstracts, and conference proceedings.

Windows Live Academic is designed to help you pinpoint what you're looking for and not waste time with irrelevant content. Sort your results by author, date, publication, and conference, and check out the preview pane before you click through to see an article's abstract and citation information.

The service is currently in beta—which means it hasn't been officially launched but is fully functional and ready to use. There are currently articles indexed in the fields of computer science, physics, electrical engineering, and related subject areas.

http://academic.live.com/

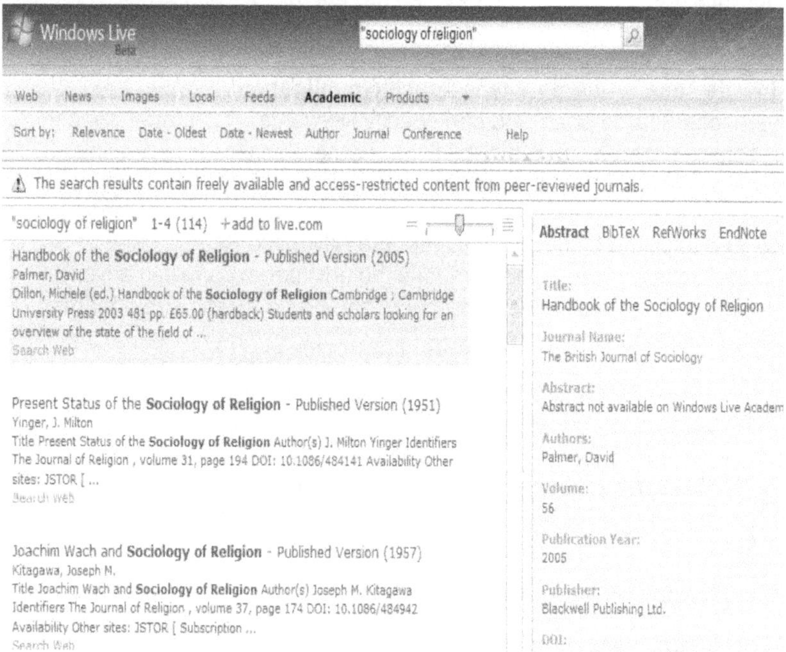

Figure 3.1 – A Windows Live Academic Search Result

Using the Wikipedia

Wikipedia, an international project that uses Wiki software to collaboratively create an encyclopedia, is becoming more and more popular. Everyone can directly edit articles and every edit is recorded. The version history of all articles is freely available and allows a multitude of examinations.

Wikipedia's founder, Jimmy Wales, has asked that college students refrain from citing Wikipedia as a source of academic research. Aside from the obvious accuracy debate, the mere nature of Wikipedia as a constantly changing open source encyclopedia goes against the purpose of academic attribution.

Wikipedia, however, provides a quick overview of topical information when conducting research. It is useful for quickly finding a broad range of opinions on a subject.

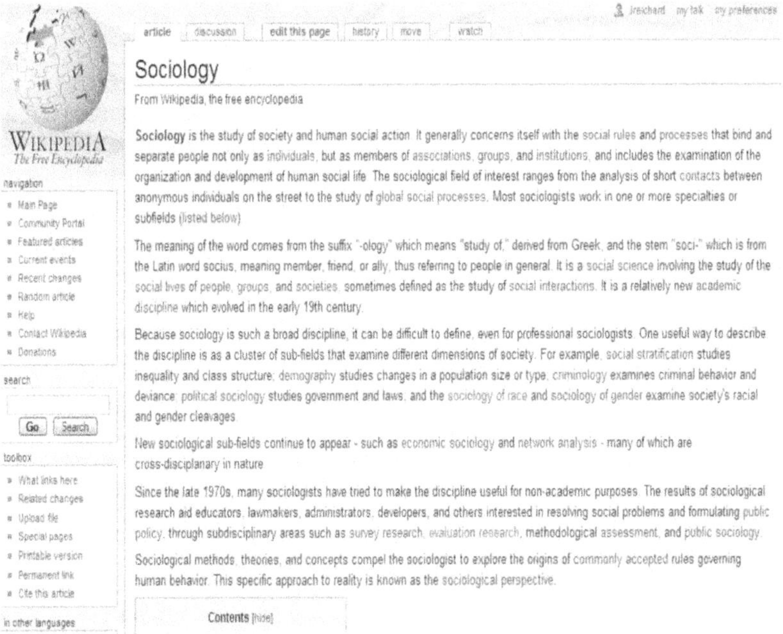

Sociology

From Wikipedia, the free encyclopedia

Sociology is the study of society and human social action It generally concerns itself with the social rules and processes that bind and separate people not only as individuals, but as members of associations, groups, and institutions, and includes the examination of the organization and development of human social life The sociological field of interest ranges from the analysis of short contacts between anonymous individuals on the street to the study of global social processes. Most sociologists work in one or more specialties or subfields (listed below)

The meaning of the word comes from the suffix "-ology" which means "study of," derived from Greek, and the stem "soci-" which is from the Latin word socius, meaning member, friend, or ally, thus referring to people in general. It is a social science involving the study of the social lives of people, groups, and societies, sometimes defined as the study of social interactions. It is a relatively new academic discipline which evolved in the early 19th century.

Because sociology is such a broad discipline, it can be difficult to define, even for professional sociologists. One useful way to describe the discipline is as a cluster of sub-fields that examine different dimensions of society. For example, social stratification studies inequality and class structure; demography studies changes in a population size or type; criminology examines criminal behavior and deviance; political sociology studies government and laws; and the sociology of race and sociology of gender examine society's racial and gender cleavages.

New sociological sub-fields continue to appear - such as economic sociology and network analysis - many of which are cross-disciplanary in nature

Since the late 1970s, many sociologists have tried to make the discipline useful for non-academic purposes. The results of sociological research aid educators, lawmakers, administrators, developers, and others interested in resolving social problems and formulating public policy, through subdisciplinary areas such as survey research, evaluation research, methodological assessment, and public sociology.

Sociological methods, theories, and concepts compel the sociologist to explore the origins of commonly accepted rules governing human behavior. This specific approach to reality is known as the sociological perspective.

Contents [hide]

Figure 4.1 – The Sociology article of Wikipedia

Evaluating Internet Resources

Finding the information you want on the Internet is only the first step. There is a vast quantity of material available, but not all of it is equally reliable and useful. As a researcher a large part of your job is not simply to find information, but to make judgments about its merit. Before you use any material you have found, you need to spend some time evaluating it for accuracy and importance. Use the following questions as a guide, but also use your own experience and skills to make a decision.

Evaluating Internet resources is not that different from evaluating other kinds of resources, many of the questions about such things as authority, bias, and currency will be of equal importance.

Who put this information here?

The source of the material might give you a clue to its reliability. A site maintained by a university or government organization might be more reliable than one maintained by a private citizen.

How old is the material?

Sometimes the age of information matters. If you need current statistics then check the age of the material you have found. As a rule of thumb, in most fields anything more than five years old is probably out-dated. But a site which deals with historical information may not need up-dating as frequently as one which is all about the latest political events. Just because information isn't regularly changed doesn't mean you shouldn't use it, but you need to be aware that your information is not necessarily the most recent.

Who wrote the information? Who is responsible for this information being here?

The status of the writer is often of considerable importance in deciding the reliability of information. You can probably assume that material written or otherwise provided by a known expert in the field is likely to be reliable. Resources provided under the auspices of a recognized institution might be considered reliable as well. But what about student pages on a university server?

Just because you have never heard of the author of the page doesn't mean that the information is inaccurate or unreliable, but it does mean that you can't take it at face value. You might have to do some cross-checking, either elsewhere on the Internet or with books or articles.

Why is this material here?

Who put the material on the Internet and why? Think about whether they might have some reason other than pure helpfulness for posting information. Many special interest groups have web pages, and while this doesn't necessarily meant the material is biased it is something you need to think about. All sorts of groups now have web pages on the Internet, and obviously all of them have a message they are trying to get across. Think about what is being said, and why the material is there.

Can I do a cross check?

Think about ways you might cross check the information you have found. You might have a look at another site with similar material, ask somebody who knows something about the topic, have a look at book on the subject. Use your own experience as well. If you have already done some research in the area you will already have some knowledge of the subject. How does this material fit in with what you already know?

Evaluating Internet Resources – A Step by Step Procedure

Purpose

Audience - Consider the intended audience of the page, based on its content, tone and style. Does this mesh with your needs?

Consider the Source - Web search engines often amass vast results, from memos to scholarly documents. Many of the resulting items will be peripheral or useless for your research

Source

- Author/producer is identifiable

- Author/producer has expertise on the subject as indicated on a credentials page. You may need to trace back in the URL (Internet address) to view a page in a higher directory with background information

- Sponsor/location of the site is appropriate to the material as shown in the URL Examples:
  - **.edu** for educational or research material
  - **.gov** for government resources
  - **.com** for commercial products or commercially-sponsored sites

- ~NAME in URL may mean a personal home page with no official sanction

- Mail-to link is offered for submission of questions or comments

Content Accuracy

- Don't take the information presented at face value

- Web sites are rarely refereed or reviewed, as are scholarly journals and books

- Look for -point of view and evidence of bias

- Source of the information should be clearly stated, whether original or borrowed from elsewhere

Comprehensiveness

- Depth of information: determine if content covers a specific time period or aspect of the topic, or strives to be comprehensive

- Use additional print and electronic sources to complement the information provided

Currency -- Look to see if the site has been updated recently, as reflected in the date on the page or is the material contained on the page is current.

Links - are relevant and appropriate. Don't assume that the linked sites are the best available. Be sure to investigate additional sites on the topic.

Style and Functionality - Site is laid out clearly and logically with well organized.

Subsections - Writing style is appropriate for the intended audience. The site is easy to navigate, including clearly labeled *Back, Home, Go To Top* icons/links and internal indexing links on lengthy pages. Do the links to remote sites all work? Is the search capability is offered if the site is extensive?

Plagiarism & Referencing Electronic Resources

What is Referencing?

Referencing (also called citing) simply means that you indicate which material is not your own and show where you got it from. Even if you have not used someone's exact words, but have rephrased their ideas you need to give your sources. The idea is that someone else reading your work should be able to recognize the difference between your work and someone else's. You need to provide them with enough information about your sources that they could find the source for themselves.

There are several different referencing systems, each subject area tends to use its own system of citations, but whatever style you choose it is important to be consistent, complete and accurate. This is matters not only for books and articles, but electronic sources as well.

Why reference?
Referencing is ensuring that you if you use someone else's words or ideas, you let your readers know where the information came from. It is okay to quote from a book, or use the ideas from someone's work in your own work, but it does mean that you need to be careful to make sure that you acknowledge where the information came from, and make it quite clear that it is not your own original ideas. The way you do that is to make sure you reference any material you use which is not your own.

Referencing World Wide Web resources:

You need to include:
* author's name (if known)
* full title of the work
* title of the complete work if applicable
* document date if known
* full URL
* date of visit.

More resources about referencing
There are several different referencing systems, each subject area tends to use its own system of citations, but whatever style you choose it is important to be consistent, complete and accurate. This matters not only for books and articles, but electronic sources as well.

APA Citation of Electronic Resources:

http://www.apastyle.org/elecref.html

Portions of this appendix were modified from the original content of the following site: http://www.sofweb.vic.edu.au/internet/research.htm

APPENDIX THREE
NON-PARAMETRIC STATISTICAL PROCEDURES

These selected non-parametric procedures are based on different assumptions. Below are described in depth each procedure briefly and literature references are noted where discussions can be found. To emphasize the importance of proper measurement, the procedures are classified under the measurement levels required for each, i.e., nominal, ordinal, and interval.

Selected Non-parametric Procedures

Type of Problem	One-Sample	Two Sample Independent	Two Sample Related	Three or More Samples Independent	Three or More Samples Related
LOCATION		Median test ORDINAL DATA Mann-Whitney test- ORDINAL DATA Tukey's quick test ORDINAL DATA Randomination test for two independent samples	Sign test- Ordinal Wilcoxon matched pairs signed ranks test-ORDINAL DATA Walsh test- INTERVAL DATA Randomixation test for matched pairs INTERVAL DATA	Extensions of Meridian ORDINAL DATA Kruskal-Wallis one-way analysis of variance ORDINAL DATA	Friedman two-way analysis of variance ORDINAL DATA
DISPERSION		Moses tests of extreme reactions ORDINAL DATA			
GOODNESS OF FIT	Chi-square one sample test- NOMINAL DATA Kolmogrov-Smirknow one sample test ORDINAL DATA	Chi-square test for two independent samples- NOMINAL DATA Kolmogorov-Smirknov two sample test ORDINAL DATA		Chi-square test for k independent samples -NOMINAL DATA	
ASSOCIATION	Spearman rank correlation coefficient ORDINAL DATA	Contingency coefficient- ORDINAL DATA		Chi-square test for k independent samples NOMINAL DATA	
REGRESSION	Fitting a regression using median ORDINAL DATA Brown - Mood test ORDINAL DATA Thell Test- ORDINAL DATA	Test for Parallelism ORDINAL DATA			
MISCEL-LANEOUS	Binominal test- NOMINAL DATA One-sample rune test ORDINAL DATA	Fisher exact test- NOMINAL DATA Wald-Wolfowitz rune test- NOMINAL DATA	Mcnemar test- NOMINAL DATA		Cochran's test- NOMINAL DATA

Statistical Procedures That May Be Used With Nominal Data

Chi-square on sample test. This chi-square test is used to estimate whether a significant difference exists between observed and expected frequencies in discrete categories. When a researcher is interested in the number of objects, subjects, or responses that fall into different categories, the chi-square test can be used to assess an attribute thus classified relative to a 'Probable" classification. For example, members of a community who are divorced may be classified by their membership in a particular church (weighted for the church membership mix of the community) to determine whether the divorce rate differs across those categories. The chi-square test is one of the most widely used non-parametric statistics. A discussion on how to use it may be found in Siegel (1956), and Daniel (1978). Broffitt and Randles (1973) discuss the power of the test.

Chi-square test for two independent samples. This chi-square test is used to test the hypothesis that two groups differ with respect to some attribute and, consequently, with respect to the relative frequency with which group elements fall into several categories. For example, we can test whether two religious groups differ in their acceptance or rejection of the opinion that miracles occur today. Or a psychiatrist might determine whether "total abstainers" less frequently seek psychiatric assistance than "heavy drinkers" The application of the procedure is illustrated by Siegel (1956). Discussions and applications of variations of this test can be found in Daniel (1978).

Chi-square test for k independent samples. This procedure is a generalization of the two-sample chi-square test.

Contingency Coefficient. This procedure measures the extent of association between two sets of attributes. It tests the hypothesis that the observed value of the measure of association in a sample could have arisen by chance in a random sample from a population in which the two attributes are not correlated. For example, an accountant may test whether the use of LIFO (last in, first out) inventory valuation procedures by firms is related to their sizes, or a researcher might use this procedure to test whether the denominational affiliation to individuals is dependent upon this procedure.

Binomial Test. This procedure is designed for use with populations that consist of two, and only two, discrete classification, e.g., male and female. It is based on identifying the expected frequencies of each class, i.e. Class one =P and Class two = 1-P. The test tells us whether it is reasonable to believe that the proportions we observe in our sample have been drawn form a population having a specified value of P. A researcher might use the binomial test to assess whether the proportion of families below the poverty-line in his region is different from that of the nation generally. both Siegel (1956) and Daniel (1978) provide the mechanics for this test.

Fisher exact test. This test is designed to be used when the measurements that concern a researcher are dichotomous and the sample is small, thirty or less. It tests the null hypothesis of no difference in the, that each category is of the population. Consequently, the Fischer exact test may be used in situation similar to those when the binomial test may be used. See Siegel (1956), Daniel (1978), and Fisher (1934,1935).

Cochran's' Q test. This procedure is an extension of the McNemar test from two related samples to k related samples. It tests whether three or more matched sets of frequencies or proportions differ significantly among themselves. for example, a study might assess the impact of various events that occur during a pastor's tenure on members' preferences for the pastor to stay or to move to another congregation. After such events as fund-raising campaigns, successful revivals, unsuccessful meetings, holidays, or political campaigns, for example., the same sample of members may be polled for a "stay" or "move" vote. The Cochran test can determine whether these events have a significant effect on the members' preferences. See Cochran (1950), McNemar (1955, pp. 232-233), Siegel (1956), and Daniel (1978).

Statistical Procedures That May Be Used With Ordinal Data
Median test. The median test is a technique for assessing whether two independent groups differ in central tendencies. It test the null hypothesis that there is no difference in the medians of the two populations from which the groups are drawn. If two populations have the same median, about half of the observations in each of the two samples can be expected to be above the common median and half below. Mood (1954) discusses the power and efficiency of the test, and Siegel (1956) and Daniel (1978) give examples of its application.

Mann-Whitney test. The Mann- Whitney test is another procedure to test whether two independent groups have been drawn from the same population based on their location parameter, i.e., medians. It is one of the most powerful of the non-parametric tests and may be used as an alternative to the parametric t test when the assumptions for that rest cannot be met. Variations of the test are discussed by Mann and

Whitney (1947), Wilcoxon (1945), and \white (1952). Details of applying the procedure are given with examples by Daniel (1978) and Siegel (1958).

Tukey's quick test. this test is based on the intuition that the less overlap in observations made on two groups, the more likely the groups differ. The procedure is simple and easily performed. It tests the hypothesis that two samples come from identical population. In a study to evaluate the influence of conformity to religious practices on self-esteem, the Tukey (1950) test might be used to assess the significance of the differences in mean scores on a self-esteem inventory of two groups (conformers and non-conformers). Neave and Granger (1968) have evaluated the power of Tukey's test. Daniel (1978) explains how to use it.

The sign-test uses plus and minus signs as data, rather than quantitative measurements. It is designed to be used with paired data where it is possible to rank the two numbers of each pair with respect to each other. Using the sign test, a researcher may establish that two conditions are different by testing the null hypothesis that the median difference between the matched scores is zero. For example, in couples with children in which one parent does not attend Sunday school regularly and the other does, a researcher may assess the influence of regular Sunday school attendance on parental guidance by soliciting statements about parental guidance to be evaluated and ranked by an expert. The sign test can be used to determine if the statements of the attendee and the non-attendee differ significantly. Mood(1954) and Walsh (1946) discuss the power/efficiency of the test. Siegel (1956) and Daniel (1978) explain its use.

Wilcoxon matched-pairs signed-rank test. When the relative magnitude of differences as well as the direction of differences is known, this more powerful sign test can be used. it weights a pair that exhibits a large difference between two conditions greater than a pair that shows a small difference. When a researcher can make the judgments of direction and "greater than," the significance of the difference between two conditions can be tested with this procedure. Thus, it can be very useful for behavioral research. For example, a Sunday school superintendent might asset that a class that teaches how the Bible applies to everyday living enhances the religious perspective of students more than does a class that simply reminds students of the Biblical narrative. After some timed and otherwise controlled exposure to the two methods, students might be ranked by using the scores on a test designed to measure religious perspective. The researcher may be sure that high scores mean more religious perspective but uncertain about how much. In this situation, the Wilcoxon matched-pairs signed rank test applies. The power and efficiency of the test is considered by Mood (1954) and Siegel (1956) and Daniel (1978) demonstrate its use. The Extension of the median Test assesses whether k independent samples were drawn from the same population or populations with the same median. For example, a pastor may be interested in the relationship of church membership longevity to the amount of participation in church activities of members. Apparently two weeks of membership is not equal to one-half of four weeks of membership in terms of the influence of the church on the lifestyle of its members. Therefore, the pastor may select several ranges of time, and classify members by these selected time frames as having more or less longevity. a measurement of church activities an then be made on the members and their groups tested for differences using the extension of the median test. The test is discussed by Cochran (1954) and Mood (1950). Examples of its application are given by Siegel (1956) and Daniel (1978).

Kruskal-Wallis one-way analysis of variance. This test uses ranks to determine if three or more independent samples are from different population. It is an extension of the Mann-Whitney test. It tests the null hypothesis that k samples come from the same population or from identical populations with respect to averages. A researcher might use the Kruskal-Wallis analysis to test the hypothesis that bishops are more authoritarian than pastors. Because some pastors aspire to become bishops, the pastors are divided into two groups (those aspiring to the bishopric and those not) to analyze the variance between the three groups. Likewise, an education board may investigate the same question as it relates to teachers and administrators. Andrews (1954) has shown that the Kruskal-Wallis test is very efficient. Kruskal and Wallis (1952) and Kruskal (1952) discuss the test. Siegel (1956) and Daniel (1978) give examples of its application.

Friedman two-way analysis of variance tests the null hypothesis that three or more related (matched) samples have been drawn from the same population. it can be used to control for extraneous variables that might otherwise control for extraneous variables that might otherwise obscure a behavior being studied by hypothesis. For example, we might be interested in evaluating teaching methods by using a standard test to measure the learning achieved by each of three different methods. We match students on each of several relevant extraneous variables, e.g., grade-point averages, age, attendance, and randomly assign these matched samples to our three treatments. The Friedman analysis can be used to determine If the treatments differ. Friedman (1937) domonstrates that this test is very close to as powerful as the most powerful three-and-more-sample parametric test, i.e., the F test. Siegel (1956) and Daniel (1978) present detailed applications and discussions of the test.

Moses tests of extreme reactions should be used when the experimental conditions can be expected to elicit extreme opposite reactions in the experimental subjects. In other words, when a test based on central tendency (differences of means, medians, etc.) would obscure rather than reveal the group differences, this test should be considered. For example, a seminary president may be interested in determining whether students who have difficulty coping with their own hostility and aggressive impulses perceive interpersonal hostility displayed by others differently than those who do not have such difficulty. we might expect those with difficulty to react to the display of interpersonal hostility by either responding extremely or by repressing their response while those without difficulty exhibit average responses. The Moses test can be used to test the null hypothesis that the span of the control group (without difficulty) and that of the experimental group (with difficulty) come from the same population. Information about the test is found in Moses (1952), Siegel (1956), and Daniel (1978).

Kolmoqorov-Smirnov one-sample test. This test determines whether the sample data can be expected to have been drawn from a population having a particular theoretical distribution. The test is based on comparing the cumulative frequency distribution of the theoretical population with the observed cumulative frequency distribution . It test the null hypothesis that there is no difference between the expected number of observations in each rank and the observed number of observations. for example, perhaps a doctor or a pastor wants to determine whether a practice or a church draws equally from all socio-economic groups of the congregation and the Kolmogorov- Smirnov test may be used to determine whether the frequencies in the various categories differ significantly. This test is more powerful than the chi-square one sample test. For additional discussion and examples of applications, see Birnbaum (1952, 1953), Siegel (1956), and Daniel (1978).

The Kolmoqorov-Smirnov two-sample test determines whether two independent sample have been drawn from the same population. In the example given above, instead of being interested in whether all socio-economic groups are represented equally in a patient group or a church, a doctor or pastor may want to determine whether the groups are being brought into the practice or congregations in the proportion that they exist in the community. In this case, socio-economic composition of the community was available. A sample of the congregation can be compared to the community sample using the Kolmogorov-Smirnov two sample test. Dixon (1954) discusses the power/efficiency of the test. Kolmogorov (1941), Smirnov (1948), Siegel (1956), and Daniel (1978) provide detailed information on the technique.

Spearman rank correlation coefficient. When an observer is interested in the degree of association between two variables in a population and when ranking of observations is possible and an observer is interested in the degree of association between two variables in a population, this procedure may be used to determine the relationship hypothesis that the calculated association does not significantly differ from zero. The calculated statistic is commonly called *rho*. Tests of association are widely used today. For example, we may intuit that the monetary contributions of members of a community service club are associated directly with their socio-economic status or their family income bracket or both. Such hunches can be investigated using Spearman rank correlation. Hotelling and Pabst (1936) tested the power/efficiency of this procedure. It is discussed by Kendall (1948a, 1948b), Siegel (1956), and Daniel (1978).

Kendall rank correlation coefficient (tau) is designed for use with the same type data and in a similar designed for use with the same type data and in a similar manner as Spearman rank correlation coefficient. However, tau can be generalized to a partial correlation coefficient. This makes it possible to control for the effects of a third variable on the two variable of direct interest to an observer. The two basic procedures are equally powerful. Further discussion: Kendall (1938, 1945, 1949), Siegel (1956), and Daniel (1978).

Kendall coefficient of concordance (W) measures the correlation among three or more sets of rankings to determine the overall agreement. The procedure is particularly useful when investigating clusters of variables. For example, a pastor desires to know whether a congregation generally agrees on the relative importance of the various doctrines of the church. To investigate this attitude, the researcher lists the prominent doctrines randomly and asks each member of a random sample of the congregation to rank the doctrines in order of importance from most to least. Kendall coefficient of concordance can be used to measure the agreement among the various rankings of the participating members. A test of significance is available. Friedman (1940), Kendall (1948), Siegel (1956), and Daniel (1978) discuss the technique and its interpretation.

Fitting a regression line using the median. When the assumption on which inferences are made using the least squares method (a parametric method) are not met, a nonparametric procedure is required for regression analysis is a method that uses the intercept and slope of a linear equation to determine the association between two variables. Daniel (1978) discusses a method of simple regression analysis described by Brown and Mood (1951) and Mood (1950) for fitting a regression line by using the Median. He also discussed the Brown-Mood method for regression line and an additional method proposed by Theil (1950) for testing the slope. The power-efficiency of this (1969, 1971) among others.

Test for the parallelism of two regression lines. Hollander's (1970) test for equality of slopes of population regression lines is a variation of the Wilcoxon matched pairs signed-ranks test. This test is useful when an investigator expects the same patterns of relationships between two same variables in two different population to hold while at the same time the scalar levels of the patterns may differ; for example, a researcher may postulate that the relationship between income and age is the same for men and women even though the expected income of women, on the average is significantly lower. the test for equality of slopes can be used to compare the regression lines and thus determine whether or not the patterns are the same. Daniel (1978) describes the procedure.

One-sample runs test. If inferences are to e made from sample data, the sample must be random. The one sample runs test is a procedure for deciding whether a sample is random. It is useful as a pretest in situations where statistical inferences will be made using either parametric or some non-parametric procedures. Mood (1940), Siegel (1956), and Daniel (1978) discuss the theory and use of the test.

The Wald-Wolfowitz runs test evaluates the null hypothesis that the data from two independent samples com from identical populations by using the number of runs present in the data. The alternative hypothesis is that the populations differ in any aspect whatsoever, e.g., location or dispersion. Consequently, this procedure is most useful when no particular parameter of the population is of primary interest. For example, in an effort to select activities that are acceptable to all, a youth camp director may want to determine whether boys and girls differ in their attitudes about particular recreational activities. From the data generated by a properly constructed questionnaire, the Wald-Wolfowitz (Moses, 1952; Smith, 1953) test may determine whether significant differences exist between the two groups. Siegel (1956) and Daniel (1978) present detailed examples of how the procedure works.

Statistical Procedures That May Be Used With Interval Data

Walsh test. When an investigator can assume that difference scores in two related samples are drawn from symmetrical populations that have means equal to their medians, the Walsh test can be used to test the null hypothesis that the median is equal to zero. This test, while almost as powerful as parametric test, does not require the assumption of a normal population. Consequently, when the assumption of normality is in doubt, it may be used as a powerful substitute for the parametric t test. For example, a teacher may want to evaluate effects of two different methods of teaching on memory. The students study five distinct concepts with each method (ten concepts total) over a period of two days, selecting the concepts at random for each teaching method and alternating the methods. At the end of the period the teacher asks the students to recall and list the ten concepts, hypothesizing that the median difference between the number of concepts recalled from each teaching method is zero and testing the hypothesis with the Walsh test. Walsh (1949) provides information on the power efficiency of the test. Siegel (1958) provides detailed procedure.

Randomization test for matched-pairs for small related samples, the randomization test can be used to obtain the exact probability under the null hypothesis associated with the occurrence of observed data. While very powerful, it is a non-parametric test, and thus does not make assumptions about normality. It can be used to test the null hypothesis that two treatments are equivalent, e.g., that there is no difference in the rate of learning by students of Sunday school classes and midweek family training classes. The test has a power-efficiency of 100 percent because it uses all of the information in the sample. See Fisher (1935) and Siegel

The Randomization test for two independent samples can be used to test the significance of the difference between the means of two small independent samples. It also uses exact probabilities and thus can avoid assumptions required for the t test. An investigator may use this procedure for determining, for example, whether the attitudes toward divorce of the pastor's counsel and the Christian education board differ. Siegel (1958) gives the procedure.

Statistical Procedures That May Be Used With Ratio Data

When ratio scale measurement is achieved, powerful and complex statistical methods, such as analysis of variance (ANOVA) analysis of covariance (ANCOVA), and discriminate analysis (DA) can be applied to the data generated by a study. As social scientists have become more and more concerned with the multiple interrelationships that exist in living systems, a class of procedures termed *multivariate methods* has been developed to make it possible for experimental and quasi-experimental designs to take into consideration these multiple interrelationships. Another class of procedures termed *time series analysis* has been developed to aid the investigation of changes in given variables over time to deal more effectively with the auto correlation inherent in process. Researchers interested in time series analysis should consult Nelson (1973) for an introduction at an applied level. Green (1978) provides an introduction to multi-variant procedures. Analysis of variance is covered by various authors including Cochran and Cox (1957) and Scheffe (1969). Neter and Wasserman (1974) apply linear statistical models to regression, analysis of variance, and experimental designs. Montgomery (1976) integrates various statistical methods with experimental design.

The Statistical Package for the Social Sciences (SPSS) and other computerized statistical packages for computers (WINKS) make it easy to use these complex statistical methods. Versions of such software are now available for microcomputer use as well. Many specific procedures are available. Only a few are mentioned here.

ANOVA. A statistical technique that makes it possible to partition the total variation between two or more experimental variables into its components, i.e., treatment effects, environmental effects, and error variances.
Discriminate Analysis can be used to classify experimental objects into groups by identifying and using certain discriminating variables.

Regression Analysis is used to find a mathematical function that can be used to predict the value of one variable from the value of another variable.

Factor Analysis can be used to reduce the number of variables while retaining most of the original information.

Source: Swanson, G.A. and Green. Hollis L. (1992)
Understanding Scientific Research: A Handbook for the Social Professions.
(Chapter 5). Oxford/ACRSS Books.

Appendix Four

Suggested Readings and References

Andersen, E.B. (1991). *The Statistical analysis of categorical data* (2nd rev. ed.). New York: Springer-Verlag.

Babbie, Earl. (2001). *The practice of social research* (9th edition) Wadswoth/Thompson Learning.

Behrens, L. (1992). *The American Experience: A Sourcebook for Critical Thinking and Writing*. Boston: Allyn and Bacon.

Bell, J. (1993). *Doing your research project: a guide for first-time researchers in education and social science* (2nd ed.). Buckingham; Philadelphia: Open University Press.

Berger, R.M., & Patchner, M.A. (1988). *Implementing the research plan: a guide for the helping professions*. Newbury Park, CA: Sage.

Berger, R.M., & Patchner, M.A. (1988). *Planning for research: a guide for the helping professions*. Newbury Park, CA: Sage.

Black, M. (1952). *Critical Thinking: An Introduction to Logic and Scientific Method*. New York: Prentice Hall.

Boeck, T. M., & Rainey, M. C. (2004). *Connections: Writing, Reading, and Critical Thinking* (2nd ed.). New York: Longman.

Bohrnstedt, G.W., & Knoke, D. (1988). *Statistics for social data analysis* (2nd ed.). Itasca, IL: F.E. Peacock Publishers.

Booth, V.(1993). *Communicating in Science: Writing a Scientific Paper and Speaking at Scientific Meetings,* 2nd ed. New York, Cambridge University Press.

Boyle, G.J., & Langley, P.D. (1989). *Elementary statistical methods for students of psychology, education and the social sciences*. Elmsford, NY: Pergamon.

Brown, S.R., & Melamed, L.E. (1990). *Experimental design and analysis*. Newbury Park, CA: Sage.

Browne, M. N., & Keeley, S. M. (2005). *Asking the Right Questions: A Guide to Critical Thinking* (7th ed.). New York: Prentice Hall.

Bryman, Alan. (2008). *Social Research Methods*. Oxford University Press.

Buckingham, Alan and Peter Saunders. (2004). *The survey methods workbook: From design to analysis*. Cambridge, UK: Polity Press.

Campbell, D. & Stanley, J. (1963). *Experimental and quasi-experimental designs for research*. Chicago, IL: Rand-McNally [The book by Campbell and Stanley (1963) is considered classic in the field of experimental design.]

Chambers H. (2000). *Effective Communication Skills for Scientific and Technical Professionals*. Cambridge MA:Perseus Books,

Cohen, J. (1988). *Statistical power analysis for the behavioral sciences* (2nd ed.). Hillsdale, N.J.: Erlbaum.

Couch, J.V. (1987). *Fundamentals of statistics for the behavioral sciences* (2nd ed.). St. Paul: West Pub. Co.

Cozby, P.C. (1993). *Methods in behavioral research* (5th ed.). Mountain View, CA: Mayfield Publishing.

Creswell, John. (2002). *Research design: Qualitative, quantitative, and mixed methods approaches.* (2nd ed). Thousand Oaks, CA: Sage Publications.

Day, R.A. (1998) *How to Write & Publish a Scientific Paper* (5th ed.). Phoenix, AZ, Oryx Press.

De Vaus, David. (2001). *Research design in social research*. Thousand Oaks, CA: Sage Publications.

DeVaus, D.A. (1995). *Surveys in social research* (4th ed.). St. Leonards, NSW: Allen & Unwin.

DeVellis R.F.(1991) *Scale Development*. Newbury Park, Sage Publications, (Applied Social Research Methods Series, v. 26)

Drew, Clifford J, Michael L. Hardman and Ann Weaver Hart. (1996). *Designing and conducting research: Inquiry in education and social science.* (2nd ed.). Boston, MA: Allyn and Bacon.

Ennis, Robert H. (1996). *Critical thinking*. Upper Saddle River, NJ: Prentice Hall.

Fink, Arlene. (2003). *The survey handbook* (2nd ed.). The Survey Kit 1. Thousand Oaks, CA: Sage Publications.

Fisher, A. (2001). *Critical Thinking*. New York: Cambridge University Press.

Fowler, F.J. (1995). *Improving survey questions: design and evaluation*. Thousand Oaks, CA: Sage.

Franzosi, Roberto.(2008). *Content Analysis*. Sage.

Gephart, R.P. (1988). *Ethnostatistics: qualitative foundations for quantitative research*. Newbury Park, CA: Sage.

Glaser, Edward M. (1941). *An Experiment in the Development of Critical Thinking*, Teacher's College, Columbia University.

Glicken, Morley D. (2003). *Social research: A simple guide*. Boston, MA: Allyn and Bacon.

Goldstein, H. (1995). *Multilevel statistical models* (2nd ed.) London: E Arnold; New York: Halstead Press.

Gorard, Stephen. (2003). *Quantitative methods in social science*. New York: Continuum.

Gray, David E. (2004). *Doing research in the real world*. London, UK: Sage Publications.

Green, Hollis Lynn (2008). *Interpreting An Author's Words*, Nashville: GlobalEdAdvancePress.

Green, Hollis L. and Swanson, G. A. (2011), *Research Methods for Problem Solvers and Critical Thinkers*, USA, GlobalEd AdvancePRESS.

Gurak, Laura J. and Lay, Mary M. (2002). *Research in Technical Communication*. Ablex Publishing

Hakim, C. (1987). *Research design: strategies and choices in the design of social research*. Boston: Allen & Unwin.

Hamby, B.W. (2007)*The Philosophy of Anything: Critical Thinking in Context*. Dubuque, Iowa. Kendall Hunt Publishing Company.

Harrison, A. F., & Bramson, R. M. (2002). *The Art of Thinking*. New York: Berley Trade.

Heikki Heiskanen; G.A. Swanson (1992) *Management Observation and Communication Theory.* Quorum Books.

Hessler, R.M. (1992). *Social research methods.* St. Paul: West Pub. Co.

Hinkle, D.E., & Wiersma, S.G.J. (1988). *Applied statistics for the behavioral sciences* (2nd ed.). Boston: Houghton Mifflin.

Homan R. (1991). *The Ethics of Social Research,* New York, NY, Longman.

Hult, C.A. (1996). *Researching and writing in the social sciences.* Boston: Allyn and Bacon.

Jones, R.A. (1996). *Research methods in the social and behavioral sciences* (2nd ed.). Sunderland, MA: Sinauer Associates.

Keren, G., & Lewis, C. (Eds.). (1993). *A Handbook for data analysis in the behavioral sciences: statistical issues.* Hillsdale, NJ: L. Erlbaum.

Kerlinger, Frank Nichols and Howard B. Lee. (1999). *Foundations of behavioral research* (4th ed.). Belmont, CA: Wadsworth.

Kincheloe, J. O., Weil, D. (2004). *Critical Thinking and Learning: An Encyclopedia for Parents and Teachers.* Westport, CT: Greenwood Press.

Koeseter, Arthur. (1967) The Act of Creation. (3rd. ed.) New York: Dell Lewis-Beck, MS, ed.(1994) *Basic Measurement.* London, SAGE.

Lyberg, L. et al. (Eds.). (1997). *Survey measurement and process quality.* New York: Wiley.

Maxim, Paul. S.(1991). *Quantitative Research Methods in the Social Science.* Oxford University Press.

May, Tim.(2001). *Social research: issues, methods and process.* Open University Press.

Maleske, Robert Thomas. (1995). *Foundations for gathering and interpreting behavioral data: An introduction to statistics.* Pacific Grove, CA: Brooks/Cole Publishing Corporation.

Mangione, T.W. (1995). *Mail surveys: improving the quality.* Thousand Oaks, CA: Sage.

May, T. (1993). Social research: issues, methods and process. Buckingham; Philadelphia: Open University Press.

204 Designing Valid Research

Miller, Delbert C., and Neil J. Salkind. (2002). *Handbook of research design and social measurement*. (6th ed.). Thousand Oaks, CA: Sage Publications.

Miller, James Grier (2008). Living Systems Theory (LST). *Systems Research and Behavioral Science*. (Volume 23 Issue), [Pages 289 -290 Special Issue:) Published Online: 23 May 2006 Copyright © 2008 John Wiley & Sons, Ltd.

Mischler, E.G. (1986). *Research interviewing: context and narrative*. Cambridge, MA: Harvard University Press.

Nachmias, C., & Nachmias, D. (1992). *Research methods in the social sciences* (4th ed.). New York: St. Martin's Press.

Nardi, Peter M. (2003). *Doing survey research: A guide to quantitative methods*. Boston, MA: Allyn and Bacon.

Neuman, W. Lawrence. (2006). *Social research methods: Qualitative and quantitative approaches*. (6 th ed.). Boston, MA: Allyn & Bacon..

Neuman, William. (2007). *Basics of social research: qualitative and quantitative approaches.* Pearson, Allen and Bacon.

Organ, T.W. (1965). *The Art of Critical Thinking*. Boston: Houghton Mifflin.

Patten, Mildred L. (2001). *Questionnaire research: A practical guide*. (2nd ed.). Los Angeles, CA: Pyrczak Publishing.

Patten, Mildred L. (2004). *Understanding research methods: An overview of the essentials*. (4 th ed.). Glendale, CA: Pyrczak Publishing.

Pilcher, D.M. (1990). *Data analysis for the helping professions: a practical guide*. Newbury Park, CA: Sage.

Punch, Keith.(2005) *An Introduction to Social Research: quantitative and qualitative approaches*. Sage.

Reid, W.J., & Smith, A.D. (1989). *Research in social work* (2nd ed.). New York: Columbia University Press.

Reid. S. (1987). *Working with statistics: an introduction to quantitative methods for social scientists*. Totowa, N.J.: Rowman & Littlefield.

Royse, D.D. (1991). *Research methods in social work*. Chicago: Nelson-Hall Publishers.

Rubens P.(1995). *Science and Technical Writing: A Manual of Style*, New York, Henry Holt & Co., Inc.

Rudinow, Joel and Vincent E. Barry (2004), *Invitation to Critical Thinking*. Belmont, CA: Wadsworth/Thomson Learning.

Ruggiero, Vincent Ryan (2002), *Becoming a Critical Thinker*. Boston: Houghton Mifflin Company.

Runyon, R.P. et al. (1996). *Fundamentals of behavioral statistics* (8th ed.). New York: McGraw-Hill.

Ruane, Janet. M. (2004). *Essentials of Research Methods: A Guide to Social Science Research.* Blackwell.

Ruszkiewicz, J., Walker, J. R., & Pemberton, M. (2002). Bookmarks: A Guide to Research and Writing (2nd ed.). New York: Longman.

Sales B. (2000) *Ethics in Research with Human Participants,* Washington DC: American Psychological Association,

Salkind N.J.(2008) *Statistics for People Who (Think They) Hate Statistics.* (3rd Edition). Sage.

Schonlau, Matthias, Ronald D. Fricker, Jr., & Marc N. Elliott. (2002). *Conducting research surveys via e-mail and the web*. Santa Monica, CA: Rand.

Seech, Zachary (2005), *Open Minds and Everyday Reasoning, 2nd Edition*. Belmont, CA: Wadsworth/Thomson Learning.

Siegel, S., & Castellan, Jr., N.J. (1988). *Nonparametric statistics for the behavioral sciences* (2nd ed.). New York: McGraw-Hill.

Simon, Julian Lincoln. (2003). *Basic research methods in social science: The art of empirical investigation*. New Brunswick, NJ: Transaction Publishers.

Singleton, Jr., R.A., Straits, B.C., & Straits, M.M. (1993). *Approaches to social research. New York: Oxford University Press.*

Skinner, B. F. (1959) *Cumulative Record*. New York: Appleton-Century-Crafts.

Spoull, N.L. (1995). *Handbook of research methods: a guide for practitioners and students in the social sciences* (2nd ed.). Metuchen, NJ: Scarecrow Press.

Sutton, C. (1987). *A Handbook of research for the helping professions.* New York: Routledge & Kegan Paul.

Swanson, G.A. (1991). *Internal auditing theory : a systems view.* [with Hugh L. Marsh.] New York : Quorum Books.

Swanson, G.A. (1992). *Management observation and communication theory.* [with Heikki Heiskanen.] New York: Quorum Books.

Swanson, G.A. (1993) Macro accounting and modern money supplies. New York: Quorum Books.

Swanson, G. A. (1998). *Measurement and interpretation in accounting. A living systems theory approach.* [with James Grier Miller.]. New York: Quorum Books.

Swanson, G.A. and Green, Hollis L. (1992). *Understanding Scientific Research: An Introductory Handbook for the Social Professions,* Oxford/ACRSS Books.

Vaughn, L. (2005). *The Power of Critical thinking: Effective Reasoning about Ordinary and Extraordinary Claims.* London: Oxford University Press.

Weinbach, R.W., & Grinnell, R.M. (1995). *Statistics for social workers* (3rd ed.). White Plains, NY: Longman.

Wilcox, R.R. (1987). *New statistical procedures for the social sciences: modern solutions to basic problems.* Hillsdale, N.J.: Erlbaum.

Williams, M., & May, T. (1996). *Introduction to the philosophy of social research.* London: University College London Press.

Yates, Simeon J. (2004). *Doing social science research.* London, UK: Sage Publications: Open University.

Online International Journals

Research methods knowledge database.
http://www.socialresearchmethods.

Resource for methods in evaluation in social research.
http://gsociology.icaap.org/methods/

Research methods and statistics arena.
http://www.researchmethodsarena.com/resources/

Quantitative and Qualitative Analysis in Social Sciences.
http://www.qass.org.uk/

Social Research Update.
http://sru.soc.surrey.ac.uk/

Survey Research Methods.
http://w4.ub.uni-konstanz.de/srm/

ABOUT THE AUTHOR...

Hollis L. Green, ThD, PhD is a Clergy-Educator with public relations and business credentials and doctorates in theology, education, and philosophy. A Distinguished Professor of Education and Social Change at the graduate level for over three decades, Dr. Green is a Diplomate in the Oxford Society of Scholars, and author of 40+ books and numerous articles. He served six years as a member of the U.S. Senate Business Advisory Board and with certified membership in several public relations societies (RPRC, PRSA, and IPRC). He served pastorates in five states, was a denominational official for 18 years, and traveled in ministry and lectured in over 100 countries.

Dr. Green was the founder of Associated Institutional Developers (AID) Ltd., (1974) an international Public Relations and Corporate Consultant Company. He was Vice-President of Luther Rice Seminary (1974-1979), and became the founding President (1981) and Chancellor (1991-2008) of Oxford Graduate School, [www.ogs.edu]. As part of a global outreach, Dr Green founded OASIS UNIVERSITY (2002) in Trinidad, W.I. [www.oasisedu.org] where he continues to lecture and teach and assist the administration as Chancellor. In 2004, he assisted in establishing Greenleaf Global Educational Foundation in Colorado to advance issues related to the current needs of society.

In addition to his other endeavors, Dr. Green launched Global Educational Advance, Inc. (2007) [www.gea-books. com or www.GlobalEdAdvance.org] to advance higher education and social change through publishing, curriculum development, instruction, library/learning resources, and

About the Author

global book distribution with 30,000 distributors in 100 countries to advance social change. His books and assisting authors in publishing are a logical outgrowth of a fifty-year ministry through education. He serves the author/publisher partnership PRESS as Corporate Chair and Co-publisher with his son, Barton. Dr. Green continues to travel, speak, teach, write books and work with authors in publishing.

~

Designing Valid Research
A Brief Study of Research Methodology

Hollis L. Green, ThD, PhD

ISBN 978-1-935434-57-3